IET CONTROL ENGINEERING SERIES 78

Frequency-Domain Control Design for High-Performance Systems

Other volumes in this series:

Volume 2	**Elevator traffic analysis, design and control, 2nd edition** G.C. Barney and S.M. dos Santos
Volume 8	**A history of control engineering, 1800–1930** S. Bennett
Volume 14	**Optimal relay and saturating control system synthesis** E.P. Ryan
Volume 18	**Applied control theory, 2nd edition** J.R. Leigh
Volume 20	**Design of modern control systems** D.J. Bell, P.A. Cook and N. Munro (Editors)
Volume 28	**Robots and automated manufacture** J. Billingsley (Editor)
Volume 32	**Multivariable control for industrial applications** J. O'Reilly (Editor)
Volume 33	**Temperature measurement and control** J.R. Leigh
Volume 34	**Singular perturbation methodology in control systems** D.S. Naidu
Volume 35	**Implementation of self-tuning controllers** K. Warwick (Editor)
Volume 37	**Industrial digital control systems, 2nd edition** K. Warwick and D. Rees (Editors)
Volume 39	**Continuous time controller design** R. Balasubramanian
Volume 40	**Deterministic control of uncertain systems** A.S.I. Zinober (Editor)
Volume 41	**Computer control of real-time processes** S. Bennett and G.S. Virk (Editors)
Volume 42	**Digital signal processing: principles, devices and applications** N.B. Jones and J. D. McK. Watson (Editors)
Volume 44	**Knowledge-based systems for industrial control** J. McGhee, M.J. Grimble and A. Mowforth (Editors)
Volume 47	**A history of control engineering, 1930–1956** S. Bennett
Volume 49	**Polynomial methods in optimal control and filtering** K.J. Hunt (Editor)
Volume 50	**Programming industrial control systems using IEC 1131-3** R.W. Lewis
Volume 51	**Advanced robotics and intelligent machines** J.O. Gray and D.G. Caldwell (Editors)
Volume 52	**Adaptive prediction and predictive control** P.P. Kanjilal
Volume 53	**Neural network applications in control** G.W. Irwin, K. Warwick and K.J. Hunt (Editors)
Volume 54	**Control engineering solutions: a practical approach** P. Albertos, R. Strietzel and N. Mort (Editors)
Volume 55	**Genetic algorithms in engineering systems** A.M.S. Zalzala and P.J. Fleming (Editors)
Volume 56	**Symbolic methods in control system analysis and design** N. Munro (Editor)
Volume 57	**Flight control systems** R.W. Pratt (Editor)
Volume 58	**Power-plant control and instrumentation** D. Lindsley
Volume 59	**Modelling control systems using IEC 61499** R. Lewis
Volume 60	**People in control: human factors in control room design** J. Noyes and M. Bransby (Editors)
Volume 61	**Nonlinear predictive control: theory and practice** B. Kouvaritakis and M. Cannon (Editors)
Volume 62	**Active sound and vibration control** M.O. Tokhi and S.M. Veres
Volume 63	**Stepping motors: a guide to theory and practice, 4th edition** P.P. Acarnley
Volume 64	**Control theory, 2nd edition** J.R. Leigh
Volume 65	**Modelling and parameter estimation of dynamic systems** J.R. Raol, G. Girija and J. Singh
Volume 66	**Variable structure systems: from principles to implementation** A. Sabanovic, L. Fridman and S. Spurgeon (Editors)
Volume 67	**Motion vision: design of compact motion sensing solution for autonomous systems** J. Kolodko and L. Vlacic
Volume 68	**Flexible robot manipulators: modelling, simulation and control** M.O. Tokhi and A.K.M. Azad (Editors)
Volume 69	**Advances in unmanned marine vehicles** G. Roberts and R. Sutton (Editors)
Volume 70	**Intelligent control systems using computational intelligence techniques** A. Ruano (Editor)
Volume 71	**Advances in cognitive systems** S. Nefti and J. Gray (Editors)
Volume 73	**Adaptive sampling with mobile WSN** K. Sreenath, M.F. Mysorewala, D.O. Popa and F.L. Lewis
Volume 74	**Eigenstructure control algorithms: applications to aircraft/rotorcraft handling qualities design** S. Srinathkumar
Volume 75	**Advanced control for constrained processes and systems** F. Garelli, R.J. Mantz and H. De Battista
Volume 76	**Developments in control theory towards glocal control** Li Qiu, Jie Chen, Tetsuya Iwasaki and Hisaya Fujioka (Editors)

Frequency-Domain Control Design for High-Performance Systems

John O'Brien

The Institution of Engineering and Technology

Published by The Institution of Engineering and Technology, London, United Kingdom

The Institution of Engineering and Technology is registered as a Charity in England & Wales (no. 211014) and Scotland (no. SC038698).

First published 2012

The Institution of Engineering and Technology
Michael Faraday House
Six Hills Way, Stevenage
Herts, SG1 2AY, United Kingdom

www.theiet.org

British Library Cataloguing in Publication Data
A catalogue record for this product is available from the British Library

ISBN 978-1-84919-481-5 (hardback)
ISBN 978-1-84919-482-2 (PDF)

Typeset in India by MPS Limited
Printed in the UK by CPI Group (UK) Ltd, Croydon, CR0 4YY

Contents

1 Justification for feedback control **1**
1.1 Tracking 2
1.1.1 Disturbance rejection 3
1.1.2 Sensitivity to parameter variation 3
1.1.3 Transient response 4
1.2 Exercises 5

2 Plant descriptions **7**
2.1 Mathematical preliminaries 8
2.2 Plant modeling in the frequency domain 9
2.2.1 Laplace transform 9
2.2.2 Transfer function/transfer matrix 10
2.2.3 Frequency response 11
2.2.4 Nonminimum phase system 13
2.2.5 Bode phase/gain relationship 13
2.2.6 Nonminimum phase lag 14
2.3 Plant modeling in the time domain 16
2.3.1 Solution of the state differential equation 17
2.3.2 Controllability and observability 19
2.3.3 Minimal state space realizations 20
2.3.4 Diagonalizing the state matrix 20
2.3.5 Transfer function from the state equation 21
2.4 Linearization 22
2.5 System identification 24
2.6 Exercises 25

3 Feedback **27**
3.1 Feedback 28
3.2 Sensitivity 29
3.3 Bode sensitivity integral 31
3.4 Bandwidth limitations 31
3.4.1 Sensor noise 31
3.4.2 Actuator limits 32
3.4.3 Plant limits – poles 32
3.4.4 Plant limits – zeros 34

3.4.5 Plant knowledge 37
3.4.6 Time delay 38
3.5 Exercises 38

4 Feedforward **41**
4.1 Command feedforward 41
4.2 Prefilter 46
4.3 Exercises 48

5 Stability **49**
5.1 Bounded-input, bounded-output stability 49
5.1.1 Marginally stable systems 50
5.1.2 BIBO stability of state equations 51
5.2 Zero input stability 52
5.2.1 Hidden modes 52
5.3 Nyquist Stability Criterion 54
5.4 Relative stability 63
5.5 Internal stability 67
5.6 Generalized Nyquist Stability Criterion 68
5.7 Gershgorin analysis 68
5.7.1 Case study: multiaxis control of a parallel robot 69
5.8 Lyapunov method 74
5.9 Direct method 77
5.10 Case study: set point control of a parallel robot 78
5.11 Kinematic set point control 78
5.11.1 Kinematic control law 80
5.11.2 Examples of kinematic set point control 80
5.11.3 Effect of mechanism singularities on the kinematic set point controller 82
5.12 Absolute stability 83
5.12.1 Circle Criterion 84
5.12.2 SISO case 85
5.12.3 Popov Criterion 86
5.13 Exercises 87

6 Feedback design – linear **91**
6.1 The Bode loop response 93
6.1.1 Shaping the response below crossover: roll-off slope in between first and second order 93
6.1.2 Shaping the loop response above crossover: high-frequency slope and the Bode step 95
6.1.3 The complete loop shape 97
6.1.4 Case study: loop shaping 98
6.2 Phase stabilization 106
6.3 Nyquist-stable system 109

6.4 Two-input, single-output control 109
6.5 Single-input, two-output control 113
6.5.1 Case study: SITO control of a ship's rudder 115
6.5.2 Case study: poor application of Nyquist-stable control 120
6.6 Exercises 122

7 Feedback design – nonlinear 127
7.1 Anti-windup 127
7.2 Nonlinear dynamic compensation 128
7.2.1 Case study: nonlinear dynamic compensator design for a vibration suppression system 129
7.3 Multipurpose nonlinear dynamic compensation 143
7.3.1 Case study: anti-windup control 145
7.3.2 Case study: nonlinear dynamic compensation for multiple saturations 148
7.4 Variable gain for SITO feedback systems 155
7.4.1 Case study: HSS rate error variable gain SITO controller 156
7.5 Exercises 160

8 References 163

Appendix: Proof of Bode sensitivity integral 165
Bibliography 171
Index 177

For Boris J. Lurie

Chapter 1
Justification for feedback control

If a PhD candidate in control theory is asked, 'in what applications should feedback be applied?', he will typically struggle to deliver a learned response. This usually is not a consequence of a lack of rigorous training in control theory. More often than not, it is the result of weaknesses in the training process itself. The academy has over the last few decades increasingly distanced itself from control applications in favor of strictly theoretical development. The result is a population of PhDs in control who is expert in elegant (if not always applicable) mathematics and crashingly ignorant in the engineering of control design. This has opened an ever-widening rift between the academic and industrial control communities. The former largely disregards the latter as primitive; the latter disregards the former as purveyors of the useless.

So, what is the answer to this most basic questions? Before it is presented, the liabilities associated with feedback control are discussed. Compare two systems with block diagrams shown in Figures 1.1 and 1.2. The first system is *open-loop*, where the output y is the result of function G acting on input u. G is a mathematical model of the *plant*. Often the output is the product Gu, as is the case when applying the Laplace Transform operator to both the system's impulse response and the input function. Methods of plant modeling are presented in Chapter 2. The second system is *closed-loop*, where the sum of the output and sensor noise n is compared to the reference input, and the difference e is input to system G_c in cascade with G. G_c is the *feedback compensator*, the result of the engineer's design to improve the performance of the system. The following are deleterious effects of the application of feedback:

1. The feedback system requires the purchase and integration of a sensor(s). This increases the cost in comparison to the open-loop system.
2. All sensors generate noise. This signal is in the feedback loop and generates an unwanted input to the plant.
3. The application of feedback can result in instability. This is discussed in detail in Chapter 5.
4. In most cases, the application of feedback reduces the gain of the system. This is discussed in detail in Chapter 6.

What does the feedback provide that justifies the problems it presents? Consider some applications where the use of feedback might be indicated.

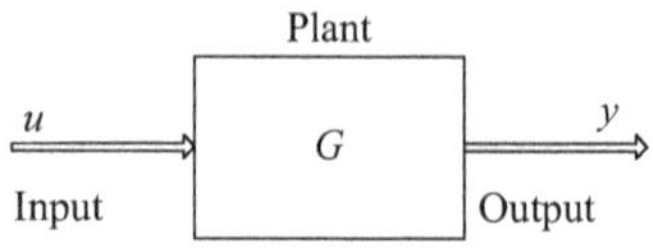

Figure 1.1 An open-loop system

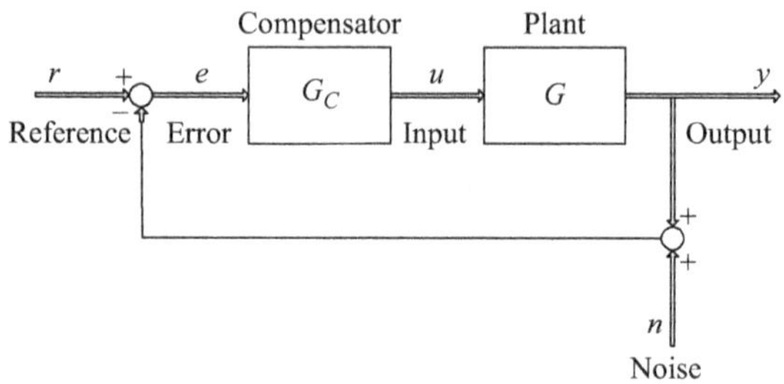

Figure 1.2 A closed-loop system

1.1 Tracking

A common requirement is that the plant output follows the input. Unless G is unity, the open-loop system will not track signal u. If the plant is a dynamic system, the response is defined in part by the equations of G. The feedback system input to the plant is a function of the tracking error $r - y$, and a correctly designed system can force the plant to track the input. So is the feedback mandatory for the tracking application? Consider the case where the plant parameters are invariant, and function G models the plant perfectly. The open-loop control G^{-1} in series with G tracks the input (Figure 1.3). So, why would one suffer the expense and risk of introducing feedback when such a straightforward solution to the tracking problem exists?

There are inherent weaknesses in this open-loop approach. G^{-1} may be an improper or unstable system. Another problem occurs in practical applications. The assumptions that G perfectly models the plant, and that the plant parameters are invariant are almost never good ones. In addition to plant parameter uncertainty that is always a consideration, such tight modeling requires expensive actuation, with well-known and invariant parameters. The feedback tracking system acts on the difference between the desired and actual output, and as such is superior to the open loop system for most applications.

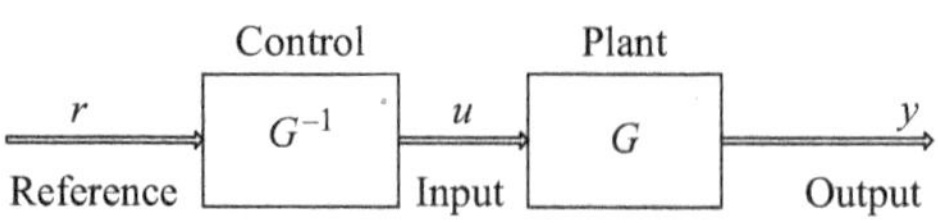

Figure 1.3 A tracking system using plant inversion

1.1.1 Disturbance rejection

The environment typically has unwanted influence on the output of the system. This can be modeled as a source d additive at either the input or output of the plant block (Figure 1.4). The effect of these *disturbances* must be attenuated. How is this achieved? Consider the case of the disturbance signal d known by the designer perfectly, and the actuator dynamics are perfectly modeled by A, which has inverse A^{-1}. Then the disturbance can be canceled by the input signal $u = -A^{-1}d$.

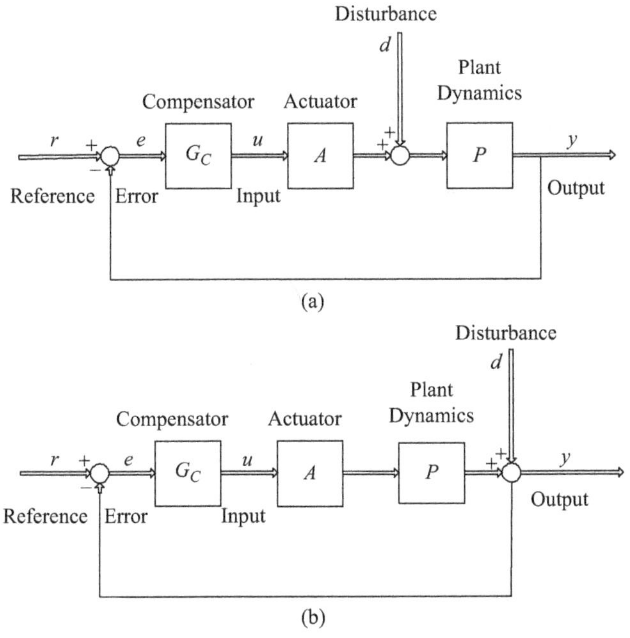

Figure 1.4 Block diagram of a control system with disturbances modeled: (a) additive at the plant input and (b) additive at the plant output

Similar to the tracking case, the open-loop approach to disturbance rejection is usually not practical. Perfect knowledge of the disturbance signal is rare. A perfect model A requires extremely precise (and expensive) actuation. The well-designed feedback control system makes the system *automatic*, tracking the reference input and rejecting the disturbance with cheaper actuation.

1.1.2 Sensitivity to parameter variation

Fatigue, temperature sensitivity and the like may change actuator and plant dynamics. For the open-loop tracking system, fixed compensator G^{-1} no longer necessarily perfectly cancels the dynamics, and performance is reduced. It will be shown that the feedback system reduces sensitivity to parameter variation.

1.1.3 Transient response

Transient response characteristics (typically related to the input step function) are considered with tracking and disturbance rejection performance. A list of common transient response (to step input) characteristics follows.

1. Rise time: the time for the system to reach a fraction of its final value, typically 0.9.
2. Overshoot: the ratio of the peak response value to its final value.
3. Settling time: the time after which the response is bounded within an interval, typically ±5 or 10% of the final value.

The plant response to input u may not be acceptable. Excessive overshoot, or lengthy settling times are examples, shown in Figures 1.5 and 1.6. Feedback can improve these characteristics, but aggressive feedback controllers, while improving the rise time, can degrade overshoot and settling time. As we will see in Chapter 6, for frequency domain feedback designs, transient response characteristics of the closed-loop system should be considered secondary to maximizing feedback. *Prefilters* are designed to improve the transient response of these systems *after* the feedback control design is completed. In this manner, time domain and frequency domain performance measures are considered separately.

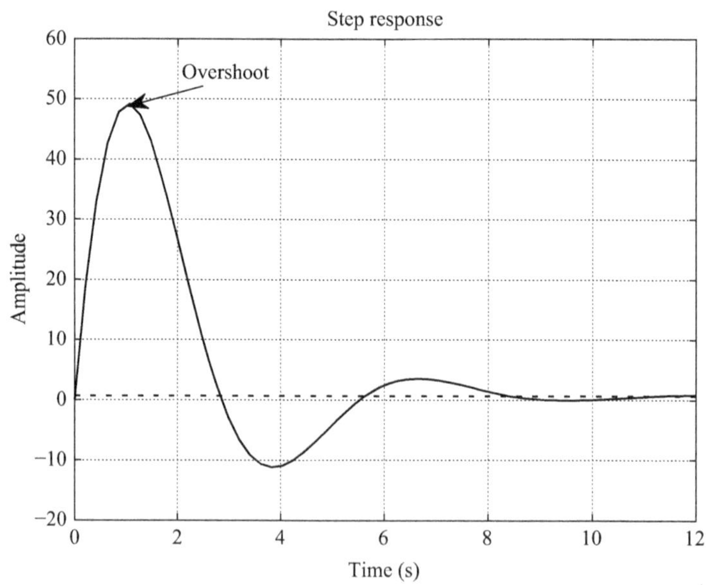

Figure 1.5 Examples of poor temporal responses: excessive overshoot

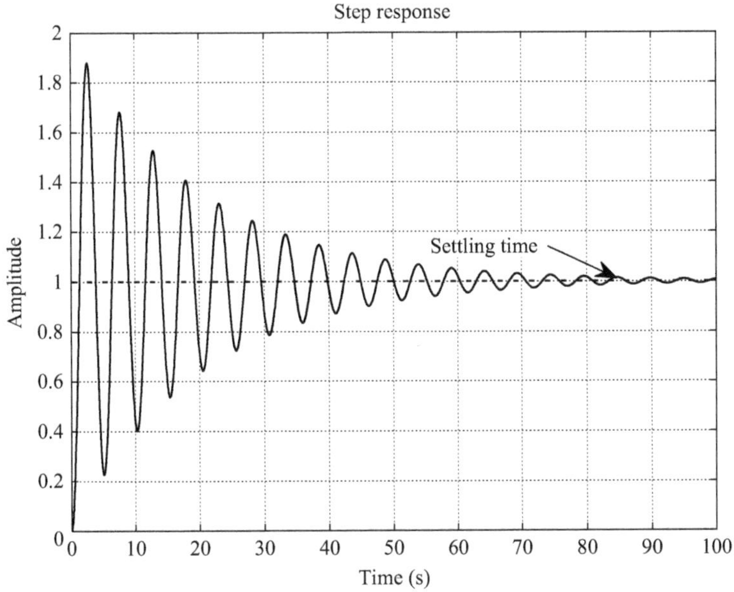

Figure 1.6 Examples of poor temporal responses: excessive settling time

1.2 Exercises

1. In what circumstances are the application of feedback indicated?
2. Explain the limitations of open-loop control.
3. Describe the conditions whereby open-loop tracking is perfect. Why is this rare in actual applications?
4. What are the disadvantages of feedback control?
5. How can the transient response of a system be improved without the application of feedback?

Chapter 2
Plant descriptions

> Understanding the plant is the most important part of control design. Everybody knows control theory ... most ignore the plant.
>
> – *Anonymous*

Accurate knowledge of plant characteristics is essential for high-performance control design. Depending on the plant, good knowledge of input mapping might be difficult to obtain. For instance, the dynamics of the plant might be nonlinear and designer only knows the system's characteristics in the close neighborhood of a particular operating point. In such a case, control is feasible only in this neighborhood. The plant might be highly resonant, but the control designer only has good knowledge of the modes up to a boundary frequency. Feedback control at frequencies higher than this is potentially destructive.

Typically, a mathematical model of the plant dynamics, consisting of a set of equations with time or frequency as the independent variable, is developed for control design purposes. Although useful from a mathematical perspective, the development of a plant model of sufficient fidelity for high-performance control can be an arduous, if not impossible, task. Simple systems with lumped parameters, such as a mass-spring-damper with a force input, are well suited for accurate model development. The three parameters of this second-order system are usually easy to find. However, it is common in real applications for the plant to have additional dynamics beyond what is described in first principles models. Although the second-order model of the mass-spring-damper captures the rigid body mode, it does not include the flexible body dynamics that the actual system might exhibit at higher frequencies, the presence of which can threaten the stability of a feedback controller if not compensated for. To augment the second-order model to include these dynamics is no longer the straightforward task of determining easily measured parameters, but a complicated modeling task, perhaps involving distributed parameters. Confidence in such high-order models is often low.

An alternative is to determine the system's response experimentally. Using the mass-spring-damper example, the response of the system to input forces can be measured by a sensor placed on the mass (e.g. an accelerometer). The input and response signals are recorded and processed so that the plant dynamics are revealed. Additional dynamics not captured by the rigid body model may be revealed without necessitating the development of a mathematical model that included them and control design can proceed taking into consideration these potential hazards. However, there are drawbacks. Sensor noise may contaminate the measurements sufficiently to degrade plant knowledge. Experimental identification of nonlinear systems might give response information accurate only for specific inputs. Indeed, a plant may be ill-suited for this approach, as its operating conditions might preclude the experiment, or it may lack stability without feedback control.

2.1 Mathematical preliminaries

A system with one input and one output is called *single-input, single-output* (SISO). SITO, TISO and MIMO are acronyms for single-input, two-output; two-input, single-output; and multiple-input, multiple-output systems, respectively. A *linear* system satisfies the superposition principle (given system operator $G(\cdot)$, $G(a_1u_1 + a_2u_2) = a_1G(u_1) + a_2G(u_2)$ for inputs u_1, u_2 and scalars a_1, a_2). A system is *time invariant* if the output does not explicitly depend on time (i.e. input $u(t)$ produces output $y(t)$, then shifted input $u(t-\tau)$ produces $y(t-\tau)$). A system that satisfies both the linearity and time invariant conditions is *linear, time invariant* or LTI. A system is *causal* if for any two signals $u_1(t) = u_2(t), \forall t \leq T$, two outputs $y_1(t), y_2(t)$ are produced that satisfy $y_1(t) = y_2(t), \forall t \leq T$.

The set of real, imaginary and complex numbers are $\Re$, $\Im$, and C, respectively. The complex number $s \in C$ is in the open right half plane (ORHP) if $Re(s) > 0$, otherwise it is in the closed left half plane (CLHP). The open left half plane (OLHP) and closed right half plane (CRHP) are defined similarly. $\Re^+$ and $\Re^-$ are the sets of positive and negative real numbers, respectively. The set of integers, positive integers (counting numbers), negative integers and nonnegative integers are Z, Z^+, Z^- and Z^{+0}. For variable s, the set of polynomials in s is $F[s] = \{p(s) : p(s) = a_ns^n + a_{n-1}s^{n-1} + \cdots + a_1s + a_0,\ a_i \in \Re,\ n \in Z^{+0}\}$. For $p(s) \in F[s]$, $p(s) = a_ns^n + a_{n-1}s^{n-1} + \cdots + a_1s + a_0$, n is the *degree* of the polynomial, $deg(p(s))$. An $n \times m$ matrix of polynomials in s is an element of set $F^{n\times m}[s]$.

The set of rational functions is $\mathbf{F}(s) = \{P(s) : P(s) = \frac{p(s)}{q(s)},\ p(s) \in \mathbf{F}[s],\ q(s) \in \mathbf{F}[s]\}$. The *relative degree* of $P(s) \in \mathbf{F}(s), P(s) = \frac{p(s)}{q(s)}$, is $deg(q(s)) - deg(p(s))$. Rational functions with nonnegative relative degrees are *proper*; those with positive relative degrees are *strictly proper*. An $n \times m$ matrix of rational functions in s is an element of $F^{n\times m}(s)$.

Given $G(s) \in \mathbf{F}(s), G(s) = \frac{n(s)}{d(s)}, s \in \{s \in C : G(s) = 0\}$ is a *zero* of $G(s)$ and $s \in \{s \in C : G(s) \to \infty\}$ is a *pole* of $G(s)$. Repeated roots of $n(s)$ and $d(s)$ must be counted as individual zeros and poles, respectively. The definitions of poles and zeros of elements of $\mathbf{F}^{n\times m}(s)$ are more complicated. Zeros of individual rational function elements are not generally zeros of the matrix $P(s) \in \mathbf{F}^{n\times m}(s)$ (there are several different types of multivariable zeros), and while the pole locations for $P(s)$ in the complex plane can be determined by finding the poles of the individual rational function elements, their multiplicity may be difficult to determine. The Smith-McMillan transformation [39] is useful in finding multivariable zeros and pole multiplicity.

Logarithmic gain is typically reported in decibels ($20 \log_{10}(x)$ dB). For example, a gain of 2 is $20 \log_{10}(2) = 6$ dB. A gain of 0.1 is $20 \log_{10}(0.1) = -20$ dB. Frequency intervals are either base 2 logarithmic, *octaves* (oct), or base 10 logarithmic, *decades* (dec).

2.2 Plant modeling in the frequency domain

Plant models may have either time or frequency as the independent variable. So-called *modern* control tends to emphasize time-domain models, whereas classical or Bode-type control usually uses frequency-domain models. Time-domain modeling is described in the sequel; however, the majority of the material in this book is focused on the frequency domain. The reason for this is rooted in the ability to quickly assess relative stability and performance when designing in the frequency domain. Although automatic control designs prevalent in modern control (e.g. linear quadratic Gaussian) lend themselves well to computer programming, it can be difficult to quantify performance and relative stability. Manual methods, like those discussed in subsequent chapters in this book, are not well suited for computer programming and require substantial insight on the part of the control designer. However, the information at a glance obtained from frequency domain plots usually more than makes up for this.

2.2.1 Laplace transform

The one-sided Laplace transform is

$$L[f(t)] = F(s) = \int_{0^-}^{\infty} f(t)e^{-st}dt \tag{2.1}$$

where $s = \sigma + j\omega$ is the Laplace variable. Given the Laplace transform $F(s)$, the function $f(t)$ can be found using the inverse Laplace transform:

$$L^{-1}[F(s)] = \frac{1}{2\pi j}\int_{\sigma-j\infty}^{\sigma+j\infty} F(s)e^{st}ds \tag{2.2}$$

$$= f(t)1^{+}(t) \tag{2.3}$$

where $1^{+}(t)$ is the unit step function (equal to 1 for $t \geq 0$, equal to 0 if $t < 0$). Table 2.1 lists properties of the Laplace transform.

Typically inverse Laplace transforms are found by decomposing the rational function $F(s)$ using partial fraction expansion. Common Laplace transforms and their inverses are shown in Table 2.2.

Table 2.1 Properties of the Laplace transform

	Time domain	Laplace domain
Linearity	$af(t) + bf(t)$	$aF(s) + bF(s)$
Time differentiation	$\frac{d^n f(t)}{dt^n}$	$s^n F(s) - s^{n-1} f(0) - s^{n-2} f'(0) - \cdots - f^{(n-1)}(0)$
Frequency differentiation	$t^n f(t)$	$(-1)^n \frac{F^n(s)}{ds^n}$
Integration	$\int_0^t f(\tau)d\tau$	$\frac{1}{s}F(s)$
Scaling	$f(at),\ a > 0$	$\frac{1}{a}F\left(\frac{s}{a}\right)$
Frequency shifting	$e^{at}f(t)$	$F(s - a)$
Time shifting	$f(t - a)1^+(t - a)$	$e^{-as}F(s)$
Convolution	$(f * g)(t)$	$F(s)G(s)$

Table 2.2 Common Laplace transforms and their inverses

Function	$f(t)$	$F(s)$
Unit impulse	$\delta(t)$	1
Unit step	$1^+(t)$	$\frac{1}{s}$
Unit ramp	$t1^+(t)$	$\frac{1}{s^2}$
Delayed unit step	$1^+(t - \tau)$	$\frac{e^{-\tau s}}{s}$
Exponential	$e^{at}1^+(t)$	$\frac{1}{s+a}$
Sine	$\sin(\omega t)1^+(t)$	$\frac{\omega}{s^2+\omega^2}$
Cosine	$\cos(\omega t)1^+(t)$	$\frac{s}{s^2+\omega^2}$
Damped sine	$e^{-at}\sin(\omega t)1^+(t)$	$\frac{\omega}{(s+a)^2+\omega^2}$
Damped cosine	$e^{-at}\cos(\omega t)1^+(t)$	$\frac{s-a}{(s+a)^2+\omega^2}$

2.2.2 Transfer function/transfer matrix

Consider the *n*th order, LTI differential equation for a SISO LTI system with zero initial conditions.

$$a_n \frac{d^n y(t)}{dt^n} + a_{n-1} \frac{d^{n-1} y(t)}{dt^{n-1}} + \cdots + a_1 \frac{dy(t)}{dt} + a_0 y(t)$$
$$= b_m \frac{d^m u(t)}{dt^m} + b_{m-1} \frac{d^{m-1} u(t)}{dt^{m-1}} + \cdots + b_1 \frac{du(t)}{dt} + b_0 u(t) \quad (2.4)$$

where $u(t)$ is the input to the system and $y(t)$ is the output.

The Laplace transform of this differential equation is straightforward using the derivative property.

$$a_n s^n Y(s) + a_{n-1} s^{n-1} Y(s) + \cdots + a_1 s Y(s) + a_0 Y(s)$$
$$= b_m s^m U(s) + b_{m-1} s^{m-1} U(s) + \cdots + b_1 s U(s) + b_0 U(s) \quad (2.5)$$

where $Y(s)$ and $U(s)$ are the Laplace transforms of the output and input signals, respectively. The *transfer function* of the system, $G(s) \in \mathbf{F}(s)$, is the ratio of Laplace transforms of the output and input:

$$G(s) = \frac{b_m s^m + b_{m-1}s^{m-1} + \cdots + b_1 s + b_0}{a_n s^n + a_{n-1}s^{n-1} + \cdots + a_1 s + a_0} \tag{2.6}$$

Given the system impulse response function, $g(t)$, the output is this function convolved with the unit impulse function, $y(t) = \int_0^t g(t-\tau)\delta(\tau)d\tau$. From the convolution property of the Laplace transform (Table 2.1), the Laplace transform of the output is $L(y(t)) = G(s)L(\delta(t))$, where $G(s)$ is the Laplace transform of the impulse response. Recall the Laplace transform of the unit impulse function is unity, so $L(y(t)) = G(s)$.

For a MIMO LTI system with zero initial conditions, the Laplace transforms of inputs $u_j(t), j = 1, 2, \ldots, m$, are mapped to the Laplace transforms of outputs $y_i(t)$, $i = 1, 2, \ldots, n$ through the $\mathbf{G}(s) \in \mathbf{F}^{n\times m}(s)$ *transfer matrix*. Element $g_{ij}(s) \in \mathbf{F}(s)$ is the Laplace transform of the ith output to the unit impulse at the jth input.

2.2.3 Frequency response

Given transfer function $G(s)$, the *frequency response* of the system is the complex function $G(j\omega) = |G(\omega)|e^{j\phi(\omega)} = a(\omega) + jb(\omega)$, $\omega > 0$, where ω is the radian frequency. The modulus, $|G(\omega)| = |a(\omega) + jb(\omega)|$, and argument, $\phi(\omega) = \tan^{-1}\left(\frac{b(\omega)}{a(\omega)}\right)$, are often plotted against log frequency in either Hz or rad/s. The log modulus (the logarithmic gain) is typically plotted in decibels ($20\log_{10}|G(\omega)|$) and the argument is plotted linearly. The combination of these plots is called the *Bode* plot. Analysis using the Laplace transform is called *frequency domain* analysis.

A logarithmic gain approximation of the Bode plot is easily established. Consider an LTI system with transfer function $G(s) = \frac{1}{\tau s+1}$, which has one pole and no finite zeros. The logarithmic gain is

$$\begin{aligned} 20\log_{10}|G(\omega)| &= 20\log_{10}\left(\frac{1}{\tau^2\omega^2+1}\right)^{\frac{1}{2}} \\ &= -10\log_{10}(\tau^2\omega^2+1) \end{aligned} \tag{2.7}$$

At $\omega \ll \frac{1}{\tau}$ (frequencies much lower than the pole frequency), $\tau^2\omega^2 + 1 \approx 1$ and the logarithmic gain is approximately 0 dB. At $\omega \gg \frac{1}{\tau}$, $\tau^2\omega^2 + 1 \approx \tau^2\omega^2$, and $20\log_{10}|G(\omega)| \approx -20\log_{10}(\tau\omega)$. At these frequencies, the logarithmic gain is decreasing. To determine the slope, consider the difference in gain over a decade, $\omega_2 = 10\omega_1$.

$$\begin{aligned} 20\log_{10}(\tau\omega_1) - 20\log_{10}(\tau\omega_2) &= -20\log_{10}\frac{\tau\omega_1}{\tau\omega_2} \\ &= -20\log_{10}\frac{\omega_1}{10\omega_1} \\ &= 20 \end{aligned}$$

The logarithmic gain drops 20 dB over a decade, or the slope is -20 dB/dec. It is clear that for each additional pole, the slope is decreased by -20 dB/dec, and that each zero increases the slope by 20 dB/dec. Over an octave (i.e. $\omega_2 = 2\omega_1$), each pole reduces the slope by -6 dB/oct (each zero increases the slope by 6 dB/oct). At high frequencies (much higher than the frequencies of the poles and zeros), the logarithmic gain slope can be approximated by the relative degree of the transfer function. For example, a system with a transfer function with five poles and two zeros will have a logarithmic slope at high frequency of -18 dB/oct (relative degree 3), and a system with a transfer function with one pole and two zeros will have a logarithmic slope at high frequency of 6 dB/oct (relative degree -1).

An approximation of the phase is also easily found. Consider a system with transfer function $G(s) = \frac{a}{s+a}$. The frequency response is $G(j\omega) = \frac{a}{j\omega+a}$ for $\omega \in \Re^{+0}$, and the phase is $\phi(\omega) = \tan^{-1}(\frac{-\omega}{a})$. Consider three frequencies referenced to the pole frequency spread across four decades $\omega = 0.01a, 0.1a, a, 10a, 100a$. At $\omega = 0.01a$, $\phi(\omega) = \tan^{-1}(-0.01) = -0.01$ rad. At $\omega = 0.1a$, $\phi(\omega) = \tan^{-1}(-0.1) = -0.0997$ rad. At $\omega = a$, $\phi(\omega) = \tan^{-1}(-1) = -\frac{\pi}{4}$ rad. At $\omega = 10a$, $\phi(\omega) = \tan^{-1}(-10) = -1.471$ rad. Finally, at $\omega = 100a$, $\phi(\omega) = \tan^{-1}(-100) = -1.561$ rad. From this, it is evident that the phase can be approximated as flat at 0 degrees from 0 to one decade below the pole frequency. In the two decades from $0.1a$ to $10a$, the phase drops approximately $\frac{\pi}{2}$ radians. At higher frequencies, the phase is approximately $-\frac{\pi}{2}$ radians. It is clear that two poles generate $-\pi$ radians at high frequency and a zero $\frac{\pi}{2}$ radians. Much like the case of determining logarithmic gain slope at high frequency, the relative degree of the transfer function allows an expeditious approximation of the phase at high frequency. For example, a system with a transfer function with five poles and two zeros will have a phase at high frequency of -270 degrees (relative degree 3), and a system with a transfer function with one pole and two zeros will have a phase at high frequency of 90 degrees (relative degree -1).

The approximations of gain and phase of the frequency response become inaccurate for systems with very light damping. For instance, the Bode plot gain approximation of $G(s) = \frac{1000}{s^2+s+100}$ would consist of a gain plot flat at 20 dB to the pole frequency of 10 rad/s, then rolling off at -40 dB/dec. The phase approximation would be flat to 1 rad/s, then drops -180 degrees over two decades to 100 rad/s. However, $|G(j10)| = 40$ dB; the approximation is off by a factor of 10 at the pole frequency. In addition, approximately 172 degrees of the 180 phase delay occurs in a two octave interval centered at the pole frequency. This is much sharper than the two decades of the approximation. For lightly damped systems, the Bode plot approximation must be augmented by calculations of gain and phase at a few points in the close neighborhood of the pole (or zero) frequency.

Care must be taken when approximating the phase at low frequency, for the existence of right half plane (RHP) poles and zeros profoundly effects the phase there.

2.2.4 Nonminimum phase system

An LTI system is *minimum phase* (MP) if it and its inverse are stable and causal. An LTI system described by $G(s) \in \mathbf{F}(s)$ is stable if all the poles have negative real parts (i.e. in the OLHP). This is discussed in Chapter 5. A system that does not satisfy the minimum phase condition is *nonminimum* phase (NMP). Limitations caused by nonminimum phase are discussed in later chapters.

2.2.5 Bode phase/gain relationship

The Bode formula provides the relationship of phase and gain for a minimum phase transfer function $G(s) \in \mathbf{F}(s)$. The phase shift at frequency ω_0 is

$$\phi(\omega_0) = \frac{1}{\pi}\int_{-\infty}^{\infty} \frac{dG}{du} \ln\left(\coth\frac{|u|}{2}\right) du \tag{2.8}$$

where $u = \ln(\frac{\omega}{\omega_0})$ and $G(\omega)$ is the modulus of the frequency response of $G(s)$. In approximating the phase associated constant magnitude slopes, it is useful to note that $\int_{-\infty}^{\infty} \ln(\coth\frac{|u|}{2})du \approx 4.93$. Consider a slope of -6 dB/oct, $\frac{dG}{du} = -\frac{\log(2)}{\log(2)} = -1$ and from (2.9), $\phi(\omega_0) \approx \frac{(-1)(4.93)}{\pi} = -1.57 \simeq -\frac{\pi}{2}$ rad. For a -12 dB/oct slope, the phase is -3.14 which is approximately $-\pi$ rad.

It is evident from the Bode phase/gain relationship that phase shift at any frequency depends on the magnitude at all frequencies. Consider a low-pass frequency response with magnitude 1 (0 dB) to a corner frequency of 1 rad/s. The modulus slope is -6 dB/oct after the corner frequency. The goal is to determine the phase at frequencies much less than the corner frequency ($\omega \ll 1$). The modulus slope $\frac{dG}{du} = -1$ at the corner frequency of 1, so the lower limit of the Bode equation is $u = \ln(\frac{1}{\omega_0}) = -\ln(\omega_0)$. Note $u = \ln(\frac{\omega}{\omega_0}) = \ln(\omega) - \ln(\omega_0)$, so $du = d\ln(\omega) = \frac{d\omega}{\omega}$. Utilizing the identity

$$\ln\left(\coth\frac{|u|}{2}\right) = \ln\left|\frac{1+\frac{\omega_0}{\omega}}{1-\frac{\omega_0}{\omega}}\right| \tag{2.9}$$

the Bode equation can be expressed as follows:

$$\phi(\omega_0) = -\frac{1}{\pi}\int_1^{\infty} \ln\left|\frac{1+\frac{\omega_0}{\omega}}{1-\frac{\omega_0}{\omega}}\right| \frac{d\omega}{\omega} \tag{2.10}$$

$$= -\frac{1}{\pi}\int_1^{\infty} \left(\ln\left|1+\frac{\omega_0}{\omega}\right| - \ln\left|1-\frac{\omega_0}{\omega}\right|\right)\frac{d\omega}{\omega} \tag{2.11}$$

If $\omega_0 \ll 1$, the integrand is approximately $|\frac{2\omega_0}{\omega}|$ and the phase is approximately

$$\phi(\omega_0) \approx -\frac{2}{\pi}\omega_0\int_1^{\infty}\frac{d\omega}{\omega^2} = -\frac{2}{\pi}\omega_0 \tag{2.12}$$

For a modulus slope of $-6n$ dB/oct ($n \in Z^+$), the phase at frequencies much below the corner frequency is approximately $-\frac{2}{\pi}n\omega_0$.

2.2.6 Nonminimum phase lag

Phase lag that is in excess of what is found using the Bode formula is *nonminimum phase lag*. RHP zeros in the transfer function generate nonminimum phase (transfer function inverse is unstable). In the time domain, such systems exhibit *inverse response*, where the initial response to the input is in the opposite direction of the final value.

Example: RHP zeros
An LTI system has transfer function $G(s) = \frac{s-10}{s^2+200s+10{,}000}$. There is one finite zero and two poles. Using the approximation of the Bode formula, the phase at very low frequency ($\omega_0 \ll 10\mathrm{rad/s}$) is approximately $-\frac{2}{\pi}\omega_0$. However, the RHP zero generates 180 degrees of phase at $\omega_0 = 0$. This is evident in the Bode plot of this system, shown in Figure 2.1. This is in contrast to the phase caused by a zero in the LHP at DC, which is 0 degrees. Figure 2.2 shows the step response of this system. Note the initial response is in the opposite direction of the final value.

It is interesting to note the phase contribution of RHP zeros at different frequencies. As seen in this example, the RHP zero contributes 180 degrees at $\omega = 0$ in contrast to 0 degrees caused by the LHP zero. As $\omega \to \infty$, both the LHP and RHP zeros contribute 90 degrees of phase. At very high frequencies relative to the zero frequency, the LHP and RHP zeros are indistinguishable from the perspective of phase. It is the phase delay of the RHP zero in the neighborhood of its frequency (transition from 180 to 90 degrees of phase contribution) that typically is most critical.

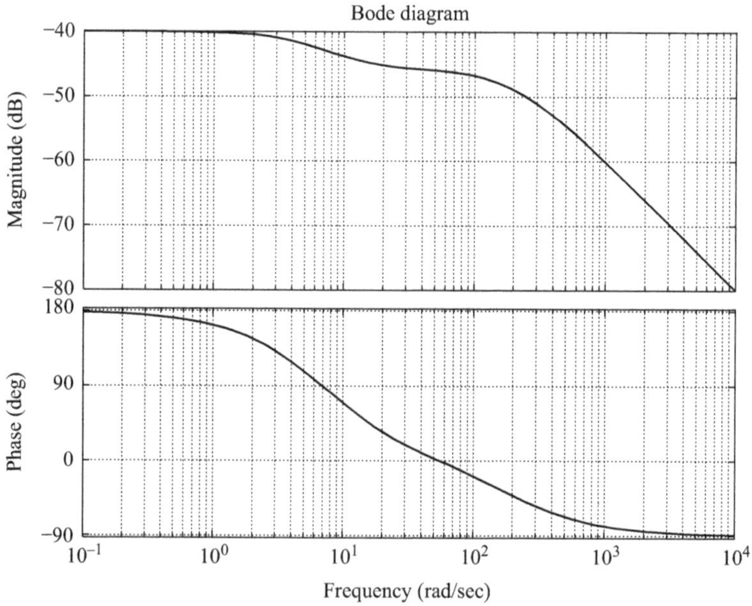

Figure 2.1 Bode plot of a nonminimum phase system

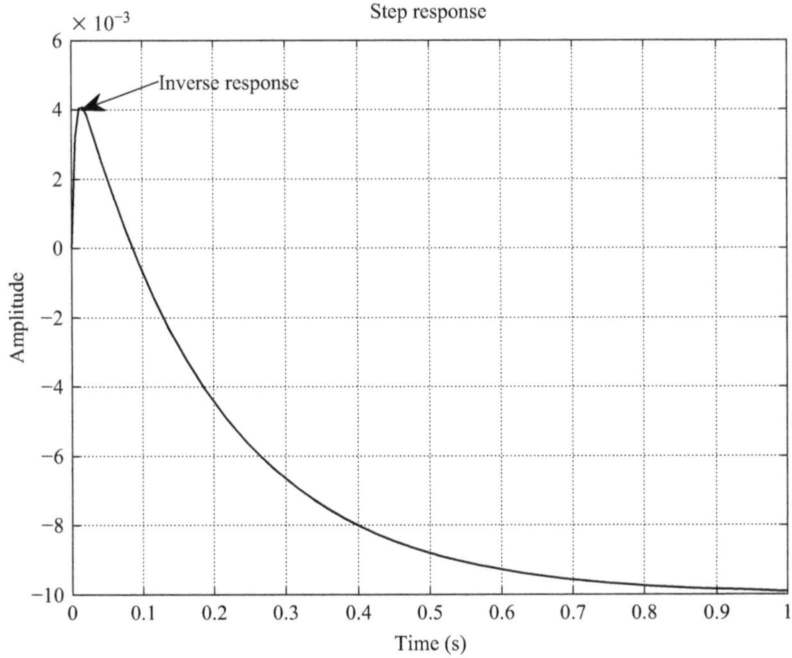

Figure 2.2 An inverse response

Transport lag $e^{-s\tau_d}$ in the loop also generates nonminimum phase delay (inverse is not causal). For a time delay of τ_d, the NMP phase delay at frequency ω is $\phi_\mathrm{delay} = \omega\tau_\mathrm{d}$. For a sampled system that waits until the end of the sample to output, $\frac{1}{\tau_d} = f_s$ is the sample rate in Hz. At the sample frequency, the time delay causes 2π radians of nonminimum phase delay. At one-tenth the sample rate, the NMP delay is 36 degrees.

Example: Time delay
The phase of the nominal system with transfer function $G(s) = \frac{100}{s^2+20s+100}$ is compared to the same system with a 0.01 s time delay in Figure 2.3. The minimum phase transfer function has a relative degree of 2 (two poles and zero finite zeros), and thus the phase will approach -180 degrees at frequencies much higher than the pole frequencies. This is seen in the solid line function. The dotted line function is the phase of the delayed system. Note that at $\omega = \frac{2\pi}{0.01}$, the excess delay is 2π. It should be noted that the plot is logarithmic and wrapped, not graphically indicating the linear relationship between time and NMP delays.

Subsequent discussions will illustrate that NMP has deleterious effects for feedback controllers. NMP feedback systems have inferior sensitivity to those that are minimum phase. Increased delay threatens stability, both in a relative and absolute sense, so there is an upper bound of acceptable NMP delay. In the case of time delay, this presents an upper bound on *frequencies* in which the controller can operate, which has a direct effect on the performance of the feedback system.

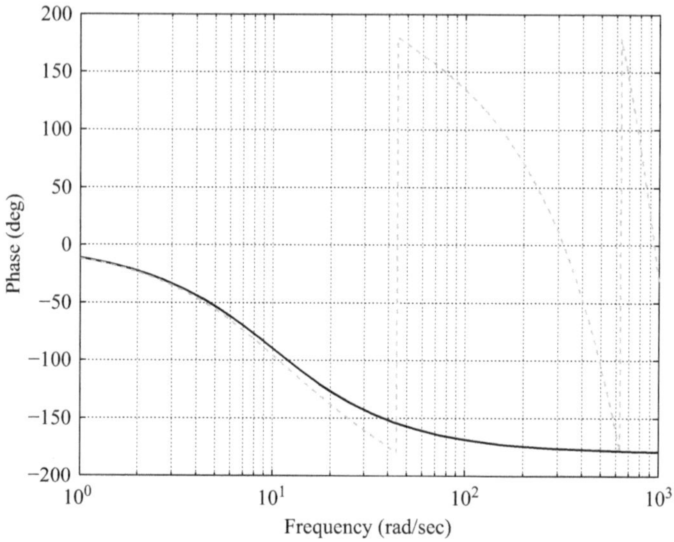

Figure 2.3 A comparison of nominal (solid line) and time-delayed (dotted line) phase plots

In the linear feedback controller design chapter, the Bode formula and quantification of NMP delay at a critical frequency will be implemented to develop a compensator shape that combines performance and robustness.

2.3 Plant modeling in the time domain

Thus far, we described linear system dynamics using the Laplace transform. Often, it is convenient to use a time-domain description. All design methods described in this book use frequency-domain representations of system dynamics and require a transformation from time-domain representations.

Definition: State

A set of variables that along with, the input functions and the equations describing the dynamics provide the future state and output of the system.

The state space realization of an n-state, LTI system with m-inputs $u(t) \in \Re^m$ and l-output $y \in \Re^l$ is the matrix quadruple (A, B, C, D), where $A \in \Re^{n\times n}$, $B \in \Re^{n\times m}$ $C \in \Re^{l\times n}$ and $D \in \Re^{l\times m}$ are the *state, input shaping, output shaping* and *input feed through* matrices, respectively. The state differential and output equations are as follows:

$$\dot{x}(t) = Ax(t) + Bu(t) \tag{2.13}$$

$$y(t) = Cx(t) + Du(t) \tag{2.14}$$

Consider a mass-spring-damper system with force applied in parallel to the spring as the input (u) and the position of the mass as the output (y). The differential equation for this second-order system is

$$M\frac{d^2y}{dt^2} + D\frac{dy}{dt} + Ky = u \tag{2.15}$$

where M, D and K are the mass, damping rate and spring stiffness, respectively. The first state, x_1, is chosen to be the mass position relative to the relaxed position. The state equations follow:

$$x_1(t) = y(t) \tag{2.16}$$

$$\dot{x}_1(t) = x_2(t) \tag{2.17}$$

$$= \frac{dy(t)}{dt} \tag{2.18}$$

$$\dot{x}_2(t) = \ddot{x}_1(t) \tag{2.19}$$

$$= \frac{d^2y(t)}{dt^2} \tag{2.20}$$

Express as a state differential equation.

$$\begin{bmatrix} \dot{x}_1 \\ \dot{x}_2 \end{bmatrix} = \begin{bmatrix} a_{11} & a_{12} \\ a_{21} & a_{22} \end{bmatrix} \begin{bmatrix} x_1 \\ x_2 \end{bmatrix} + \begin{bmatrix} b_1 \\ b_2 \end{bmatrix} u \tag{2.21}$$

$$= \begin{bmatrix} 0 & 1 \\ -\frac{K}{M} & -\frac{D}{M} \end{bmatrix} \begin{bmatrix} x_1 \\ x_2 \end{bmatrix} + \begin{bmatrix} 0 \\ \frac{1}{M} \end{bmatrix} u \tag{2.22}$$

The input is the force applied to the mass $u = f$. The position of the mass is the output $y = [1 \quad 0]\begin{bmatrix} x_1 \\ x_2 \end{bmatrix}$.

It is noted that while transfer function models of LTI systems (assuming the polynomials are coprime) are unique, state space models are not. In addition to an arbitrary order of selected states, the state matrix may be of greater dimension than the order of the system, and the state space realization is referred to as not minimal. This is discussed in more detail in Chapter 5.

2.3.1 Solution of the state differential equation

The solution of the state differential equation is now found. Consider first the zero input state differential equation.

$$\dot{x}(t) = Ax(t) \tag{2.23}$$

$$x(t) = e^{At}x(0) \tag{2.24}$$

$$e^{At} = I + At + \frac{A^2t^2}{2!} + \frac{A^3t^3}{3!} + \cdots + \frac{A^kt^k}{k!} + \cdots \tag{2.25}$$

Take the derivative.

$$\frac{d}{dt}e^{At} = A + A^2t + \frac{A^3t^2}{2!} + \cdots + \frac{A^kt^{k-1}}{(k-1)!} + \cdots \tag{2.26}$$

$$= A\left(I + At + \frac{A^2t^2}{2!} + \frac{A^3t^3}{3!} + \cdots + \frac{A^kt^k}{k!} + \cdots\right) \tag{2.27}$$

$$= Ae^{At} \tag{2.28}$$

$$= e^{At}A \tag{2.29}$$

If $x = e^{At}x(0)$, then $\dot{x} = \frac{d}{dt}e^{At}x(0) = Ae^{At}x(0) = Ax$. So $x = e^{At}x(0)$ is a solution to the state differential equation.

Consider the LTI system driven by input u.

$$\dot{x} = Ax + Bu \tag{2.30}$$

Premultiply by e^{-At} and note $\frac{d}{dt}e^{-At}x = e^{-At}\dot{x} - e^{-At}Ax$.

$$e^{-At}\dot{x} = e^{-At}Ax + e^{-At}Bu \tag{2.31}$$

$$e^{-At}\dot{x} - e^{-At}Ax = e^{-At}Bu \tag{2.32}$$

$$\frac{d}{dt}e^{-At}x = e^{-At}Bu \tag{2.33}$$

Integrate both sides of this equation.

$$\int_0^t de^{-A\tau}x(\tau) = \int_0^t e^{-A\tau}Bu(\tau)d\tau \tag{2.34}$$

$$e^{-At}x(t) - e^{-A(0)}x(0) = \int_0^t e^{-A\tau}Bu(\tau)d\tau \tag{2.35}$$

$$x(t) = e^{At}x(0) + e^{At}\int_0^t e^{-A\tau}Bu(\tau)d\tau \tag{2.36}$$

$$= e^{At}x(0) + \int_0^t e^{A(t-\tau)}Bu(\tau)d\tau \tag{2.37}$$

Now express the output equation as a function of the state transition matrix, e^{At}.

$$y(t) = Cx(t) + Du(t) \tag{2.38}$$

$$= Ce^{At}x(0) + C\int_0^t e^{A(t-\tau)}Bu(\tau)d\tau + Du(t) \tag{2.39}$$

Example. The zero input response of an LTI system is desired. The state matrix is $A = [-3\ 1;1\ -6]$ and the output shaping matrix is $C = [1\ 1]$. The initial state is $x_0 = [1;0]$.

The solution of the state differential equation is presented in Figures 2.4 and 2.5.

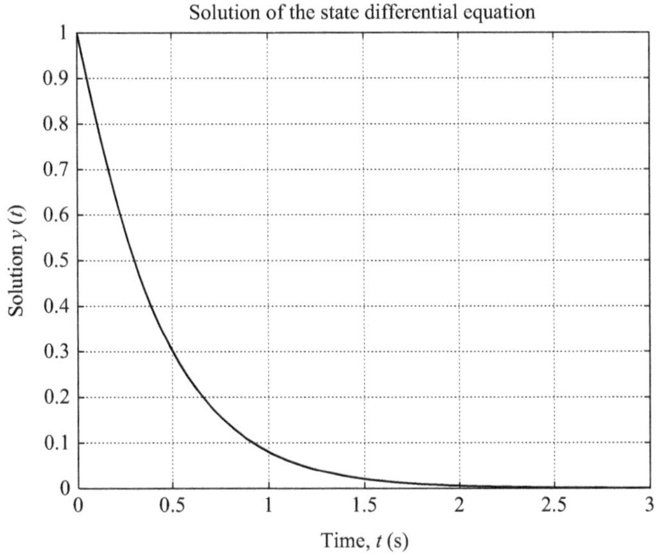

Figure 2.4 A solution of the state differential equation

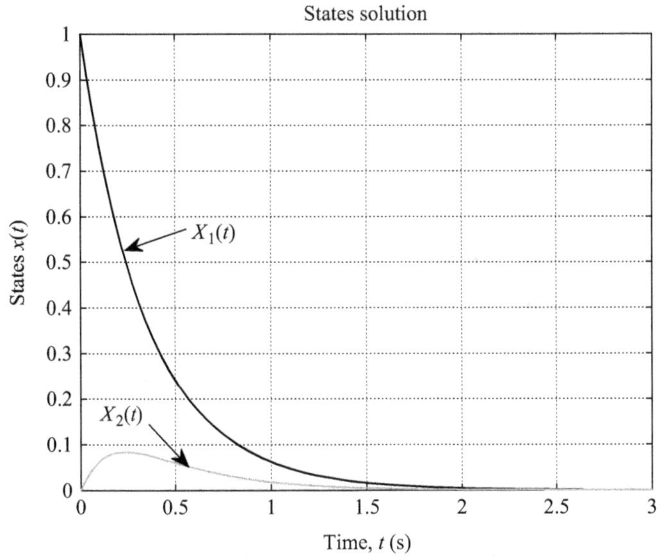

Figure 2.5 States

2.3.2 Controllability and observability

For a state-space realization (A, B, C, D), the matrix pair (A, B) is *controllable* if for any initial state x_0 and any final state x_1, there exists an input that transfers x_0 to x_1 in finite time. It can be shown that this is equivalent to matrix

$\int_0^t e^{A(t-\tau)} BB^T e^{A^T(t-\tau)} d\tau$ being nonsingular for any $t > 0$, or the controllability matrix $C = [B \quad AB \quad A^2B \quad \ldots \quad A^{n-1}B]$ is full row rank. The eigenvalues (real and conjugate pairs) of a controllable system can be placed in arbitrary locations in the complex plane with the appropriate control input.

For a state space realization (A, B, C, D), the matrix pair (A, C) is *observable* if for any unknown initial state x_0, there exists a finite time $t_1 > 0$ such that the knowledge of the input and the output over the time interval $[0, t_1]$ suffices to determine uniquely the initial state. It can be shown that this is equivalent to matrix $\int_0^t e^{A(t-\tau)} C^T C e^{A^T(t-\tau)} d\tau$ being nonsingular for any $t > 0$, or the observability matrix $O = [C^T \quad A^TC^T \quad (A^2)^TC^T \quad \ldots \quad (A^{n-1})^TC^T]^T$ is full column rank. The states of an observable system can be estimated with the output, input and the state space model of the system.

State space descriptions of linear systems are not unique. It is possible that the dimension of state matrix A is greater than the smallest possible, and the eigenvalues of A are not poles of the transfer function. This state space realization is *nonminimal* and possibly has hidden unstable modes.

2.3.3 Minimal state space realizations

Definition: Minimal state space realization

A state space realization (A, B, C, D) is minimal if A is the smallest possible dimension.

For a minimal state space realization, all eigenvalues of A are poles of the transfer matrix $G(s)$. A state space realization (A, B, C, D) is minimal if and only if (A, B) is controllable and (A, C) is observable (see Reference 3).

2.3.4 Diagonalizing the state matrix

A transform is introduced to diagonalize the $n \times n$ state matrix, A, in (2.14). The independent variable is dropped.

$$x = Mq \tag{2.40}$$

$$\dot{x} = M\dot{q} = AMq + Bu \tag{2.41}$$

$$y = CMq + Du \tag{2.42}$$

Premultiply by the inverse of transformation matrix M.

$$\dot{q} = M^{-1}AM_q + M^{-1}Bu \tag{2.43}$$

To decouple the states, $M^{-1}AM = \Lambda$ must be diagonal. Elements λ_i, $i = 1, 2, \ldots, n$, along the diagonal of Λ are the *eigenvalues* of A. Columns of M (referred to as the *modal* or *eigenvector* matrix) x_i, $i = 1, 2, \ldots, n$, are the *eigenvectors* of A. The relationship between the ith eigenvalue and eigenvector is

$$Ax_i = \lambda_i x_i \tag{2.44}$$

$$(A - \lambda_i I)x_i = 0 \tag{2.45}$$

This has a nontrivial solution for x_i if and only if $det(A - \lambda I) = 0$. If the roots of this determinant are distinct, the $n-$ eigenvectors can be found. If the nullity of $A - \lambda_i I$ is equal to the multiplicity of the ith eigenvalue, independent eigenvectors can be found for these eigenvalues. If the nullity is less than the eigenvalue multiplicity, diagonalization of the state matrix is not possible; however, a block diagonal form is possible using the Jordan transformation (see Reference 3).

The diagonalization of the state matrix A allows the expeditious analysis of the natural response of the system. Indeed, for minimal state space realizations, the eigenvalues of the state matrix are the poles of the transfer function.

2.3.5 *Transfer function from the state equation*

Consider a strictly proper SISO LTI system with state equation (2.14) ($y(t)$ and $u(t)$ are scalar functions, D is the zero matrix). We take the Laplace transform of the state equation.

$$sX(s) = AX(s) + BU(s) \tag{2.46}$$

$$(sI - A)X(s) = BU(s) \tag{2.47}$$

$$X(s) = (sI - A)^{-1}BU(s) \tag{2.48}$$

$$Y(s) = C(sI - A)^{-1}BU(s) \tag{2.49}$$

The transfer function is the ratio of Laplace transforms of the output and input, $G(s) = \frac{Y(s)}{U(s)} = C(sI - A)^{-1}BU(s)$. As $(sI - A)^{-1} = \frac{1}{det(sI-A)} Adj(sI - A)$ where $Adj(\cdot)$ is the adjoint matrix, the roots of $det(sI - A)$ are in the denominator of elements of $G(s)$. These are the eigenvalues of A. Depending on common roots, these may or may not also be poles of the system.

Example

$$\dot{\vec{x}} = \begin{bmatrix} 1 & 4 \\ 7 & 10 \end{bmatrix} \vec{x} + \begin{bmatrix} 0 \\ 1 \end{bmatrix} u \tag{2.50}$$

$$y = [1 \quad 1] \rightarrow \vec{x} \tag{2.51}$$

$$G(s) = \frac{Y(s)}{U(s)} \tag{2.52}$$

$$= C(sI - A)^{-1}B \tag{2.53}$$

$$= [1 \quad 1]\begin{bmatrix} s-1 & -4 \\ -7 & s-10 \end{bmatrix}^{-1}\begin{bmatrix} 0 \\ 1 \end{bmatrix} \tag{2.54}$$

$$= [1 \quad 1]\begin{bmatrix} \dfrac{s-10}{s^2-11s-18} & \dfrac{4}{s^2-11s-18} \\ \dfrac{7}{s^2-11s-18} & \dfrac{s-1}{s^2-11s-18} \end{bmatrix}^{-1}\begin{bmatrix} 0 \\ 1 \end{bmatrix} \tag{2.55}$$

$$= \frac{4}{s^2-11s-18} + \frac{s-1}{s^2-11s-18} \tag{2.56}$$

$$= \frac{s+3}{s^2-11s-18} \tag{2.57}$$

Note $C = \begin{bmatrix} 0 & 4 \\ 1 & 10 \end{bmatrix}$ and $O = \begin{bmatrix} 1 & 1 \\ 8 & 14 \end{bmatrix}$ are both rank 2 (the state space realization is minimal). The eigenvalues are the solution of $|\lambda I - A| = 0$ or $\lambda^2 - 11\lambda - 18 = 0$, which are the poles of $G(s)$.

2.4 Linearization

Linear control theory obviously applies to linear plants, and unfortunately most real-world systems do not satisfy the superposition condition. The dynamics of these systems can be approximated in the closed neighborhood of selected operating points. Consider a nonlinear system description $y(t) = g(x(t))$. The Taylor expansion about operating point x_0 is

$$g(x) = g(x_0) + \frac{dg}{dx}\bigg|_{x=x_0}(x - x_0) + \frac{d^2g}{dx^2}\bigg|_{x=x_0}\frac{(x-x_0)^2}{2!} + \cdots \tag{2.58}$$

If the increment $h = |x - x_0|$ is sufficiently small, the terms of order higher than 1 in the Taylor expansion are small in comparison to the first two terms, and the dynamics may be approximated as

$$g(x) = g(x_0) + \frac{dg}{dx_{x=x_0}}(x - x_0) \tag{2.59}$$

The second term is simply the product of slope $(\frac{dg}{dx}|_{x=x_0})$ and run $(x - x_0)$. Consider a two-state, nonlinear system $\dot{x}_1 = g_1(x_1, x_2), \dot{x}_2 = g_2(x_1, x_2)$, where f_1 and f_2 are continuously differentiable. Point $x_1 = p_1$, $x_2 = p_2$ is an *equilibrium point* (i.e. $\dot{x}_1 = \dot{x}_2 = 0$). The Taylor expansion about this point is

$$\dot{x}_1 = g_1(p_1, p_2) + a_{11}(x_2 - p_1) + a_{12}(x_2 - p_2) + HOT \tag{2.60}$$

$$\dot{x}_2 = g_2(p_1, p_2) + a_{21}(x_1 - p_1) + a_{22}(x_2 - p_2) + HOT \tag{2.61}$$

where $a_{ij} = \frac{dg_i}{dx_j}$, and *HOT* (high-order terms) are terms in the expansion of order higher than unity. The state trajectories in the neighborhood of (p_1, p_2) are $x_1 = x_1 - p_1$ and $x_2 = x_2 - p_2$. If $\|[x_1 \quad x_2]^T\|$ is sufficiently small, then the trajectories are approximated by

$$\begin{bmatrix} \dot{x}_1 \\ \dot{x}_2 \end{bmatrix} = \begin{bmatrix} a_{11} & a_{12} \\ a_{21} & a_{22} \end{bmatrix} \begin{bmatrix} x_1 \\ x_2 \end{bmatrix} = Ax \tag{2.62}$$

The matrix first derivative, A, is called the Jacobian matrix.

Example: Linearization

Consider a nonlinear system with state equations $\dot{x}_1 = -x_2, \dot{x}_2 = 3\sin(x_1) + x_2^2 + 3u$. The input is u and the output is state x_1. A transfer function approximating the system's dynamics in the close neighborhood of the equilibrium point (0, 0) is required. The Jacobian matrix is

$$\begin{bmatrix} 0 & -1 \\ 3 & 0 \end{bmatrix} \tag{2.63}$$

So, the linearized state space equation is

$$\begin{bmatrix} \dot{x}_1 \\ \dot{x}_2 \end{bmatrix} = \begin{bmatrix} 0 & -1 \\ 3 & 0 \end{bmatrix} \begin{bmatrix} x_1 \\ x_2 \end{bmatrix} + \begin{bmatrix} 0 \\ 3 \end{bmatrix} u = Ax + Bu \tag{2.64}$$

$$y = [1 \quad 0] = Cx \tag{2.65}$$

The transfer function is $C[sI - A]^{-1}B = \frac{-3}{s^2+3}$. Figure 2.6 compares the response of the actual and approximated system to the input $u(t) = 0.01 \sin(t)$.

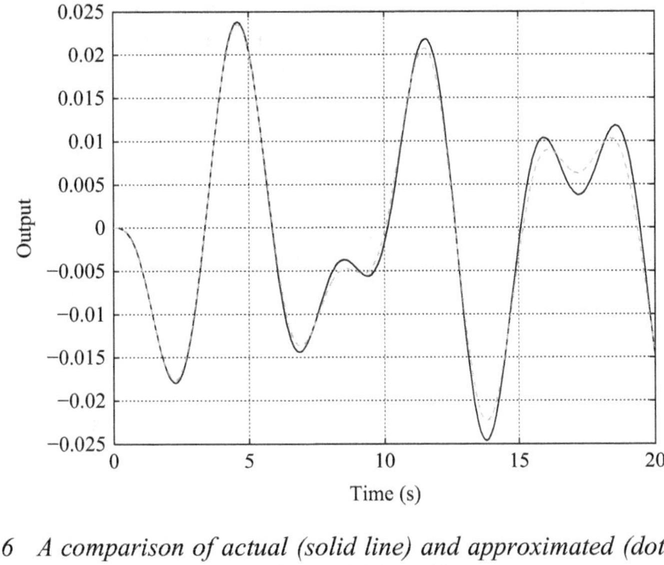

Figure 2.6 A comparison of actual (solid line) and approximated (dotted line) system responses to u(t) = 0.01 sin(t)

2.5 System identification

Models of dynamic systems can be segregated into three classes. The *white box* consists of a model developed using first principles. This type of model is useful for low-order systems (e.g. mass-spring-damper) but is difficult to develop for complex systems with many parameters. A *black box* does not use a priori model, instead uses statistical models to build a system model using measured data. This is particularly useful when modeling complicated systems with many parameters, such as those with many flexible-body modes. Developing accurate white box models of these types of systems can be very difficult. A *gray box* model consists of a combination of white and black, incorporating knowledge of the system dynamics and measured data.

The *Fourier analyzer* is a very useful tool in the development of black box models of complex linear systems. These are digital spectrum analyzers that provide through averaging sampled time responses frequency response functions, power spectral densities (PSDs), and correlation and coherence functions. These frequency response measurements can either be curve-fitted to find a rational function description, or the data can be used directly for loop shaping and stability analysis. The coherence function indicates the quality of the frequency response. At frequencies where the coherence is near unity, the response is primarily a result of the input signal. At frequencies where the coherence is substantially less than unity, the system may be unresponsive and/or the response may be driven excessively by sensor noise and the system identification may be insufficient for high-quality control design.

Example: Black box modeling with a Fourier analyzer

The system to be controlled is a single-axis vibration suppression system consisting of a voice coil actuator in parallel with a helical spring and a laser position sensor. Performance is important, so the bandwidth must be as high as possible. As such, a model of the rigid-body mode only is not sufficient for this application, as any flexible-body modes threaten the stability of an aggressive feedback system. A white box model of sufficient fidelity is not available, so a black box model is found using a Fourier analyzer. Figure 2.7 shows the coherence, gain and phase plots for this system. The coherence is near unity in the decade 100–1000 rad/s (except at the frequencies of the zeros at 50 and 90 rad/s) and at the pole frequencies in the interval 600–900 rad/s. Confidence in the model at these frequencies is high, and there is little concern if the loop gain is near unity in these frequency intervals. There is concern if the crossover frequency is in the interval from 2000 to 5000 rads/s as the coherence is low, and the knowledge of the plant phase is not very good. The low coherence at beyond 20,000 rad/s is due to the plant being unresponsive at those frequencies. A control design for this plant will be discussed in detail in a case study.

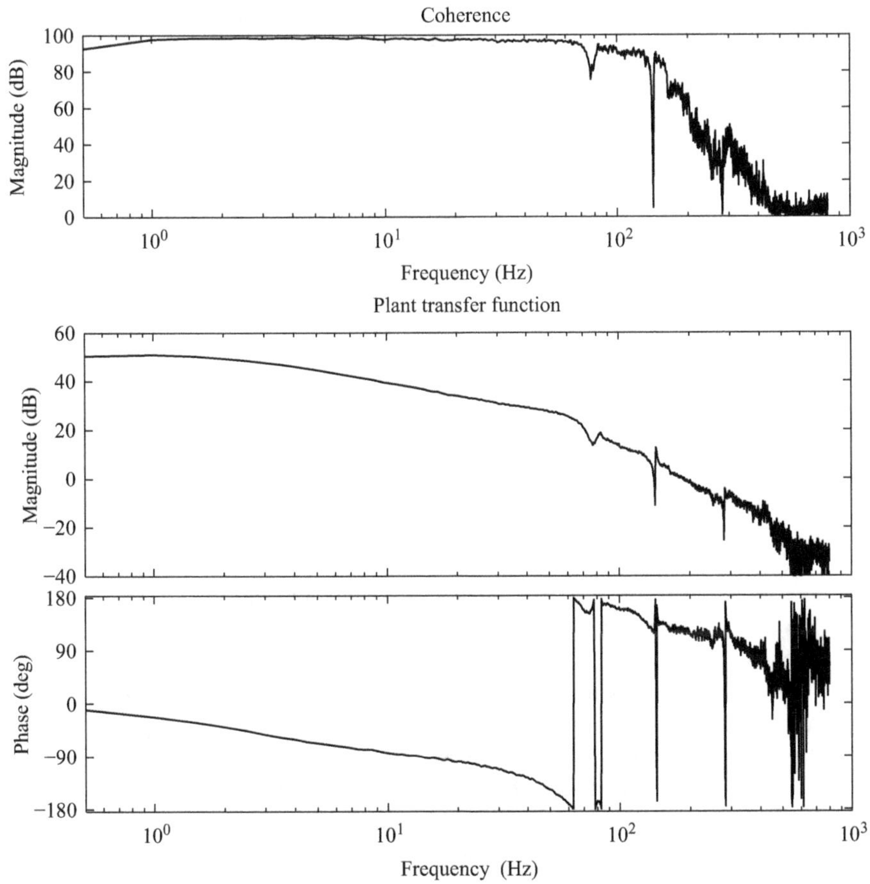

Figure 2.7 Experimentally acquired frequency response and coherence functions for a single-axis vibration suppression system

2.6 Exercises

1. The impulse response of an LTI SISO system is $g(t) = (15 + e^{-3t})1^+$. Find the transfer function $G(s)$ of this system.
2. The step response of an LTI SISO system is $g_s(t) = (t + 3t^2 + e^{5t})1^+$. Find the transfer function $G(s)$ of this system.
3. The impulse response of an LTI SISO system is $g(t) = 10(\sin(t))1^+$. Find the transfer function $G(s)$ of this system.
4. An LTI SISO system has a transfer function $G(s) = \frac{s+5}{s^2+11s+10}$. Find the response of this system to input $10e^{-5t}$.

5. What is the impulse response of the LTI SISO system $G(s) = \frac{10}{s^2+5s+100}$?
6. Sketch an approximation of the Bode plot of $G(s) = \frac{100}{s+10}$.
7. Sketch an approximation of the Bode plot of $G(s) = 1000\frac{s+5}{s^2+15s+50}$.
8. Sketch an approximation of the Bode plot of $G(s) = 1000\frac{s+1}{s^2+1s+100}$. Be sure to augment the approximation in the neighborhood of the lightly damped poles.
9. Sketch an approximation of the Bode plot of $G(s) = 1000\frac{s-1}{s^2+1s+100}$. Is the system minimum phase? Why or why not?
10. A coprime transfer function has four LHP zeros and one RHP zero. It has six LHP poles and one RHP pole. What is the slope of the frequency response at frequencies much higher than the highest frequency pole or zero? What is the phase?
11. The transfer function has a flat gain to the corner frequency 100 rad/s. The roll-off is 3rd order (−18 dB/oct). What is the phase (approximately) at 0.1 rad/s?
12. A digital sensor has a 10 ms time delay. How much phase delay is generated by this at 5 Hz?
13. The plant equation is $\frac{y^2(t)}{dt^2} + 10\frac{y(t)}{dt} = 3\frac{u(t)}{dt}$. Find a state space equation for this system. Also find the transfer function for this system.
14. For the previous system, the initial state is $x(0) = [2 \quad 0]$. The input is 0. Plot the state to 10 s.
15. Explain the conditions for which state space realization (A, B, C, D) is minimal.
16. A coprime transfer function has six zeros and eight poles. What are the dimensions of the matrices of the minimal state space realization (A, B, C, D)? What is D?
17. Find the eigenvalues and eigenvectors of state matrix $A = \begin{bmatrix} -4 & 3 \\ 1 & 20 \end{bmatrix}$. What are the poles of the system if the state space realization is minimal?
18. Linearize the system $\dot{x}_1 = x_2, \dot{x}_2 = 3e^{x_1} + x_2^2$ about the origin.

Chapter 3
Feedback

In the world of feedback control, it pays to be negative.

– *Anonymous*

Consider the block diagram of a single-input, single-output (SISO) feedback system shown in Figure 3.1. The plant and feedback compensators are modeled by the rational functions $G(s)$ and $C(s)$, respectively. Exogenous inputs $r(s)$, $d(s)$ and $n(s)$ are the reference, disturbance and sensor noise signals, and $y(s)$ is the plant response. $T(s) = C(s)G(s)$ is the *return ratio*, sometimes referred to as the *open-loop gain* or the *loop transmission function.* $F(s) = 1 + T(s)$ is the *return difference* and its magnitude $|F(s)| = |1 + T(s)|$ is the *feedback*. Although they share the same name, this function should not be confused with the feedback signal from the plant output to the compensator input. The response to reference input is $\frac{y(s)}{r(s)} = \frac{C(s)G(s)}{1+C(s)G(s)} = \frac{T(s)}{F(s)}$. The response to the disturbance is $\frac{y(s)}{d(s)} = \frac{G(s)}{1+C(s)G(s)} = \frac{G(s)}{F(s)}$. For good tracking, it is desired that $|\frac{T(s)}{F(s)}|$ be as close to the unity as possible. For good disturbance rejection, it is desired that $|\frac{G(s)}{F(s)}|$ be as small as possible.

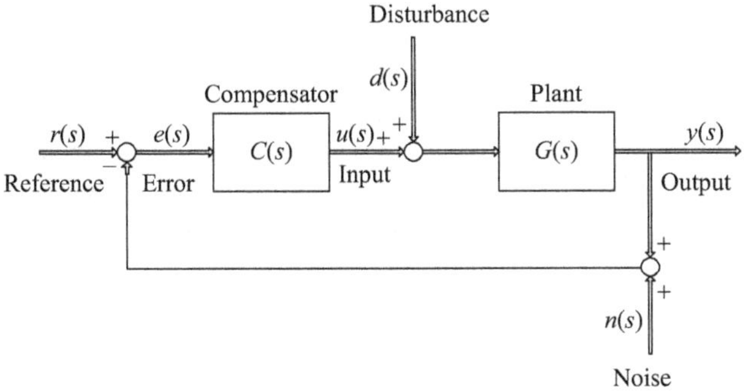

Figure 3.1 A SISO feedback system

3.1 Feedback

Definition: Bandwidth
For a low pass loop transmission and $|T(j\omega_b)| = 1$, ω_b is the bandwidth (alternatively 0 dB crossover frequency).

Definition: Functional bandwidth
For a low pass loop transmission and $|T(j\omega)|$ is approximately constant for $\omega < \omega_f$, ω_f is the functional bandwidth.

Definition: Negative feedback
Feedback is negative at frequencies where $|F(s)| > 1$. It is noted that many textbooks refer to negative feedback as a result of the negative sign on the reference/feedback summing junction. While loop phase is critical, there is nothing special about the sign at this summing junction. For instance, if a -1 block is placed in the feedback path, then the summing junction could have plus signs at both inputs with no change to the closed-loop system.

In contrast, the definition of negative feedback introduced here has a significant meaning. Note the effect of negative feedback on the response to disturbance function. In the open loop, the Laplace transform of the response of the plant to an input disturbance is $G(s)d(s)$. In the closed loop, the response is $\frac{G(s)}{F(s)}d(s)$. The response of the system to disturbance is reduced (compared to the open-loop response) over the frequencies where feedback is negative. The term 'negative' has its roots in early feedback amplifier design, where the effect of feedback reduces the high open-loop gain (negative logarithmic gain).

Definition: Large feedback
Feedback is large at frequencies where $|F(s)| \gg 1$. Clearly, it is advantageous to have large feedback, as the disturbance rejection is good at frequencies where this is true. As will be shown later, it is not possible to have large feedback at all frequencies and care must be taken in design to maximize the feedback in frequencies where disturbance rejection is most critical.

Note that at frequencies where there is large feedback, $|F(s)| = |1 + T(s)| \simeq |T(s)|$. So the loop transmission function modulus is a good approximation of the feedback where feedback is large. This approximation is used to show that large feedback improves tracking performance. Over the frequencies where feedback is large, the ratio of Laplace transforms of the response to the reference input is $\frac{y(s)}{r(s)} = \frac{T(s)}{F(s)} \simeq \frac{T(s)}{T(s)} = 1$, and the output tracks the reference input.

Feedback is the magnitude of the return difference; the phase of this complex function is ignored. In Chapter 5, it is shown that sufficient phase margins must be maintained for stability. So, why is it acceptable to consider a real function like feedback? Phase can (locally) be ignored at frequencies where feedback is large. For example, consider two systems with large feedback at some frequency ω_1 where $T_1(j\omega_1) = 1000 + j0$ and $T_2(j\omega_1) = -1000 + j0$. Although the phase of the loop transmission functions differs by 180 degrees, the feedback is nearly identical (1001 and 999). Now consider two systems at some frequency ω_2: $T_1(j\omega_2) = 1.01 + j0$ and $T_2(j\omega_2) = -1.01 + j0$. The phase again differs by 180 degrees, but note the difference in feedback: $F_1(\omega_2) = 2.01$ and $F_2(\omega_2) = 0.01$. Although not large, the feedback of system 1 at ω_2 is negative and disturbances will be reduced by approximately 6 dB at this frequency. The second system will *amplify* disturbances at ω_2 by 40 dB! This illustrates how critical the shaping of $T(s)$ is in the vicinity of the unit circle and provides an introduction to *positive feedback.*

Definition: Positive feedback
Feedback is positive where $|F(s)| < 1$. This has the opposite effect on disturbance rejection as negative feedback, as the response $\frac{G(s)}{F(s)}d(s)$ is now of greater magnitude than the open loop. Clearly, this is not a desirable characteristic. So, why not design the control system with a goal of having a wide interval of frequencies where the feedback is negative (preferably large), and no positive feedback? There is a relationship between positive and negative feedback presented later that precludes this design approach. As opposed to attempting to eliminate positive feedback, the more useful approach is to introduce positive feedback carefully at frequencies where there is low disturbance energy. This is discussed in detail later.

3.2 Sensitivity

The effect of feedback on the sensitivity of the closed-loop system to (small) plant parameter variations is now investigated. G in the reference-output equation is treated as a variable to determine this sensitivity.

$$S = \frac{d\frac{y}{r}}{dG}\frac{G}{\frac{y}{r}} \tag{3.1}$$

The first term in this product is the derivative of the reference-output equation with respect to variable G.

$$\frac{d\frac{y}{r}}{dG} = \frac{d}{dG}\left(\frac{CG}{1+CG}\right) \tag{3.2}$$

$$= \frac{C}{(1+CG)^2} \tag{3.3}$$

The sensitivity is

$$S = \frac{C}{(1+CG)^2}\frac{G(1+CG)}{CG} \tag{3.4}$$

$$= \frac{1}{1+CG} \tag{3.5}$$

$$= \frac{1}{1+T} \tag{3.6}$$

It is evident that the closed-loop system is insensitive to plant parameter variations at frequencies where the feedback is large. It should be noted that the parameter variations considered here do not include those that significantly change the plant characteristics (e.g. the inclusion or extraction of poles and zeros), rather those that effect small changes in the plant gain. The following definitions are from this result.

Definition: Sensitivity
Sensitivity is the inverse of the feedback, $S(s) := \frac{1}{1+T(s)}$. The plot of the magnitude of the sensitivity function shows the boundary between positive and negative feedback for stable loops, the horizontal line indicating 0 dB. Below this line, the feedback is negative, above it positive.

Definition: Complementary sensitivity
For sensitivity function $S(s)$, the complementary sensitivity function is defined as $M(s) := 1 - S(s)$. Note $M(s) = \frac{T(s)}{1+T(s)}$, which is the closed-loop function for unity feedback, and $S(s) + M(s) = 1$.

It is clear that control systems have several desirable characteristics at frequencies where feedback is large: good reference tracking, good disturbance rejection, and an insensitivity to small parameter variations in the plant. The last characteristic has a cost benefit. Consider a control application in which actuators from two different manufacturers satisfy the requirements of the problem (e.g. sufficient stroke, force, velocity, etc.): the first actuator is precise, expensive and somewhat fragile; the second actuator is somewhat less precise than the first, but less expensive and more rugged. What is the effect of the modulus of the frequency response of the second actuator varying slightly on the closed-loop performance, and how does this compare to the performance using the first actuator? If the controller is designed so that feedback is not large, the sensitivity is not small, and the reference tracking of the controller using the second actuator will be inferior to that using the first. The engineer is compelled to purchase the more expensive actuator and hopes that its fragility does not result in failure on deployment. However, if the controller is designed so that there is large feedback, the system is insensitive to the variations in the actuator gain, and the tracking performance using the second actuator is comparable to that using the first. The engineer may purchase the less expensive actuator and have greater confidence in system reliability.

If one desires good performance, *as much feedback should be applied at critical frequencies as is feasible*. The advantages of control systems that have large feedback include good tracking performance, good disturbance rejection and the potential for a lower cost solution to the actuator selection problem. The disadvantage is that the design of large feedback controllers is more complicated, requiring higher-order compensation than commonly used linear controllers (e.g. proportional-integral-derivative (PID)) and perhaps nonlinear compensation. This disadvantage, however, costs almost nothing. The increased compensator order results either in a few more operational amplifiers in the analog case, or a few more terms in a difference equation in the digital case. On the other hand, a few decibels of performance improvement may be the difference between the system performing to required specifications or failing to (military) or dominating the competition or languishing behind (commercial).

Having established the positive aspects of feedback, practical limitations to the available quantity are now presented.

3.3 Bode sensitivity integral

For a system with greater than first-order roll-off and N_p open-loop right half plane (RHP) poles p_i, $i = 1, 2, \ldots N_p$, the sensitivity of the closed-loop system satisfies the following (see Appendix A):

$$\pi \sum_{i=1}^{N_p} Re[p_i] = \int_0^{\infty} \log|S(j\omega)|d\omega \tag{3.7}$$

This relationship gives great insight into the limitations of control performance for systems with greater than first-order roll-off. It is noted that for low-pass loop transmission functions that roll off at first order have no positive feedback (the T-plane plot lies in the first and fourth quadrants). For stable systems (open-loop) with greater than first-order roll-off (quite typical for actual systems where the response drops sharply at high frequency), the integral of log sensitivity with positive feedback is equal to that of negative feedback. As negative feedback is increased over some intervals of frequencies, positive feedback is increased at other frequencies. From a design perspective, the compensator must be designed so that there is small disturbance power at frequencies where there is positive feedback. It is evident from (3.7) that unstable systems will have more positive feedback (from an log integral perspective) than negative feedback, and this difference is proportional to how far into the RHP these poles are. This lost performance is expended in stabilizing the unstable open-loop system.

3.4 Bandwidth limitations

Clearly, large feedback F has many advantages. The loop transmission function $T(s)$ is typically low-pass or bandpass, so increasing the crossover frequency ω_c (simultaneously reducing the first crossover frequency for bandpass $T(s)$) increases feedback. Unfortunately, arbitrarily high crossover frequencies are not possible. Some causes of bandwidth limitation follow.

3.4.1 Sensor noise

The block diagram in Figure 3.2 shows a feedback system with a linear block $H(s)$ in the feedback path and an exogenous input added to the sensor signal, $n(s)$. The frequency response of H indicates at what frequencies the sensor is responsive, and n is the sensor noise signal. The loop transmission of this system is $T(s) = G(s)C(s)H(s)$. The response of the closed-loop system to the sensor noise is as follows:

$$y(s) = \frac{-G(s)C(s)H(s)}{1 + G(s)C(s)H(s)} n(s) \tag{3.8}$$

$$= \frac{-T(s)}{1 + T(s)} n(s) \tag{3.9}$$

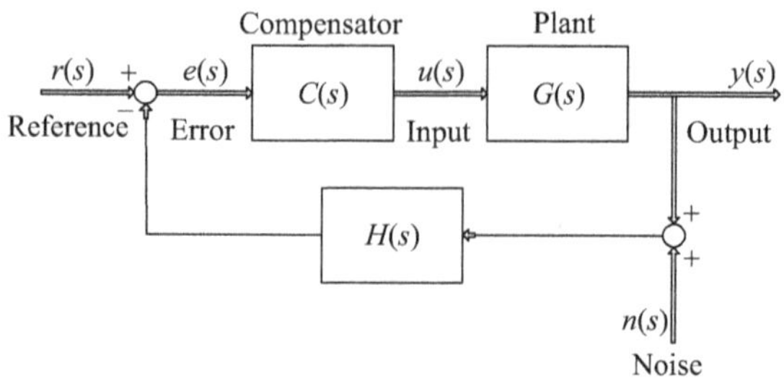

Figure 3.2 A feedback system with sensor signal added

At frequencies where feedback is large, $y(s) \simeq -n(s)$ with no other exogenous inputs. The system tracks the sensor noise, a very bad characteristic.

The sensitivity of the system to variations in H is the following:

$$S_H(s) = \frac{G(s)C(s)H(s)}{1 + G(s)C(s)H(s)} \tag{3.10}$$

$$= \frac{T(s)}{1 + T(s)} \tag{3.11}$$

At frequencies where the feedback is large, the system sensitivity to sensor variations is nearly 1, affecting the reference tracking directly. These two results comprise a compelling feature of the sensor in comparison with the actuator. Large feedback makes the system insensitive to small variations in the actuator, yet makes the system respond 1-to-1 to sensor parameter variations. Large feedback attenuates plant disturbances, yet passes sensor noise directly to the output. So while good control design (large feedback) allows some flexibility in actuator selection, the sensor used in these closed-loop systems must be high quality (invariant H and low noise power in the desired bandwidth).

3.4.2 Actuator limits

All actuators when driven to their physical limits response present a feedback limitation. Actuators typically have a limited frequency response, beyond which the output drops off. This is modeled as a low-pass filter in the forward path of the control loop, the phase lag of which must be considered in the loop shaping for adequate margins of stability.

Actuators driven beyond their amplitude limits introduce a saturation nonlinearity in the forward path. The presence of this saturation can cause instability, especially when the control design is aggressive. This is discussed in detail in Chapter 7.

3.4.3 Plant limits – poles

Many plants have dynamics that limit available bandwidth. Consider two linear, minimum phase SISO plants with frequency responses shown in Figures 3.3 and 3.4. It will be shown in Chapter 5 that at frequency ω_b where $|T(j\omega_b)| = 1$, the loop

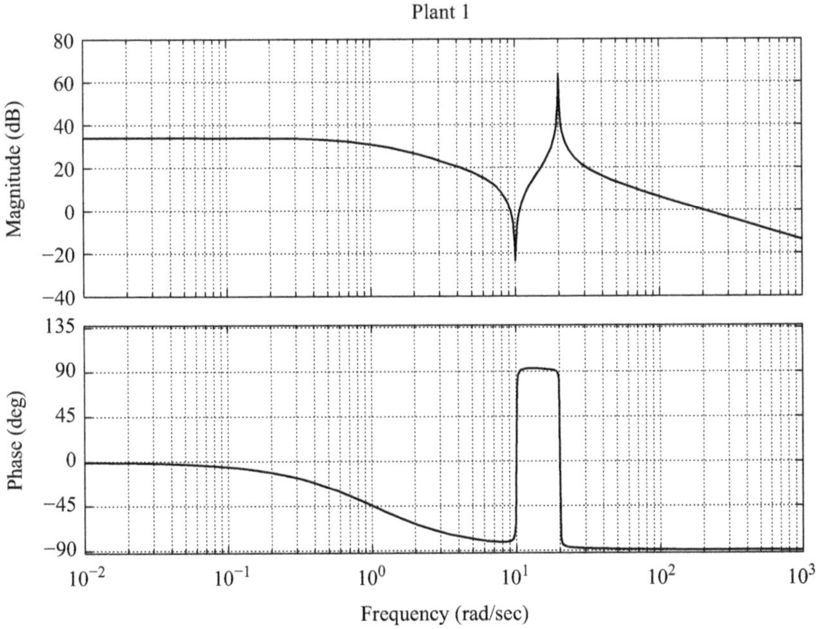

Figure 3.3 Plant 1 with alternating poles and zeros

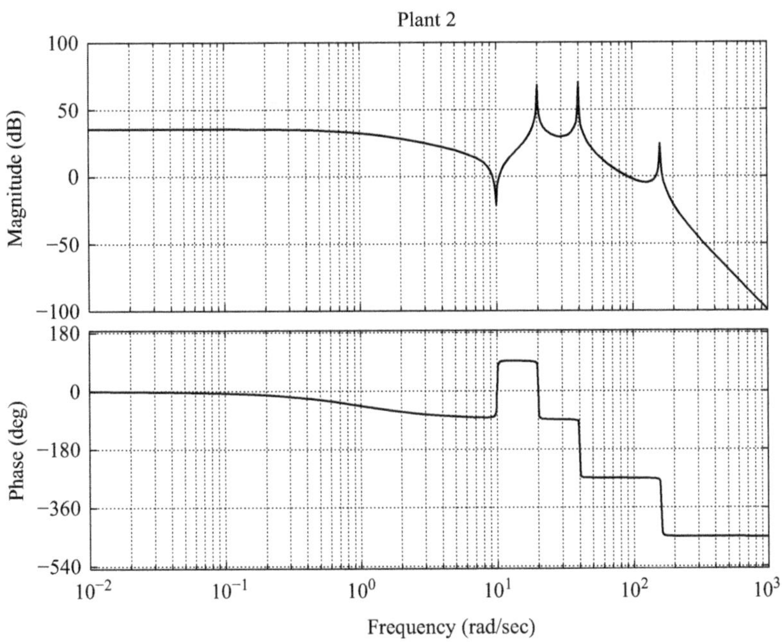

Figure 3.4 Plant 2 with adjacent poles

transmission phase angle $arg(T(j\omega_b))$ must be greater than -180 degrees. The two plants are analyzed with this constraint in mind. The first plant has only a real pole at 1 rad/s, followed by a conjugate pair of zeros at 10 rad/s and a conjugate pair of poles at 20 rad/s. The dynamics of this system do not alone restrict bandwidth. Indeed, the compensator could be nothing more than a constant, and the phase and gain stability margins are bounded above by -90 degrees and infinity, respectively. Of course, this is only feasible in the unlikely condition that there are no other limits to feedback present. The second plant breaks at the same frequency as the first but has lightly damped conjugate adjacent pairs of poles at higher frequencies. Consider the phase of this system's frequency response. The first zeros and poles occur in alternating pairs. This provides 180 degrees of phase advance in the frequency intervals between the zeros and poles. These do not present a problem in this case as the phase advances from -90 degrees (due to the real pole) to 90 degrees over the frequency interval between the zeros and poles and the Nyquist locus (discussed in Chapter 5) does not cross the negative real axis.

The situation changes at higher frequency where there are two adjacent conjugate pole pairs. The phase drops 180 degrees at each of these pole frequencies, clearly making the application of negative feedback at these frequencies a risky venture. Misguided attempts at pole-zero cancelation can have dire consequences indeed. Consider the following compensator design for the second plant. The bandwidth of the regulator is to be 200 rad/s.

$$C_1(s) = \frac{k_1\left(s^2 + 0.1s + 40^2\right)\left(s^2 + s + 160^2\right)}{\left(s^2 + 10{,}000s + 10{,}000^2\right)\left(s^2 + 10{,}000s + 20{,}000^2\right)} \tag{3.12}$$

The zeros of C_1 cancel the plant poles at 40 and 160 rad/s perfectly, and k_1 is selected so that the 0-dB crossover is 200 rad/s. The poles at very high frequency make proper the compensator transfer function. The loop transmission of Figure 3.5 shows a final 0-dB crossover of 200 rad/s. The response of this system to a random disturbance added at the plant input is shown in Figure 3.6.

A problem arises when the plant is not known perfectly, which is a very common problem with such resonant systems. To allow for some uncertainty in the target poles, the zeros are a bit wider.

$$C_2(s) = \frac{k_2\left(s^2 + 5s + 40^2\right)\left(s^2 + 5s + 165^2\right)}{\left(s^2 + 10{,}000s + 10,000^2\right)\left(s^2 + 10{,}000s + 20,000^2\right)} \tag{3.13}$$

However, even small errors in the frequency of the second pole (165 rad/s versus 160 rad/s) result in instability, as seen in the Bode plot shown in Figure 3.7, and the closed-loop response to disturbance in Figure 3.8.

3.4.4 Plant limits – zeros

Consider two-loop transmission functions shown in Figure 3.9. Note that the moduli are identical, but the arguments differ. The first system has a zero at $s = -5$, whereas the second has a zero at $s = 5$. Both systems have repeated poles

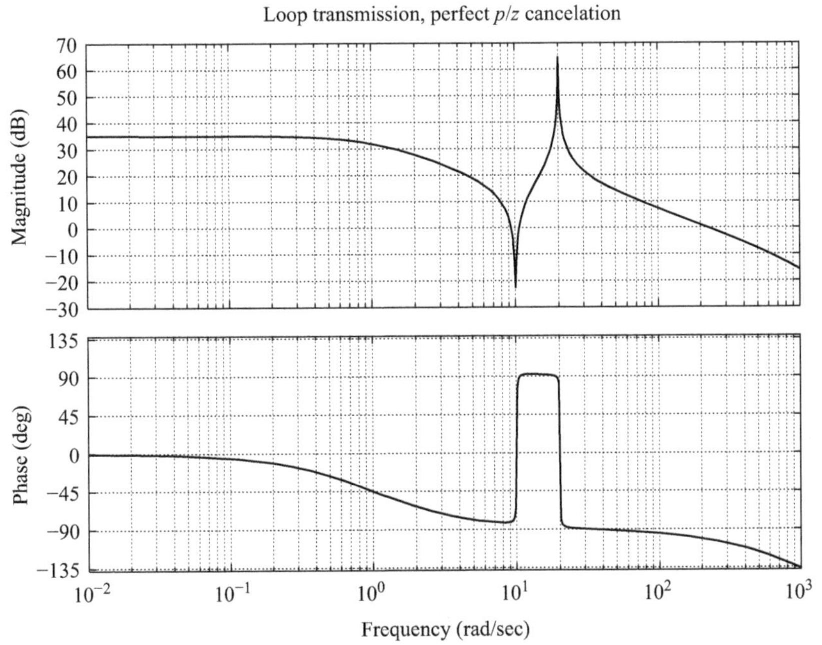

Figure 3.5 Loop transmission with perfect pole/zero cancelation

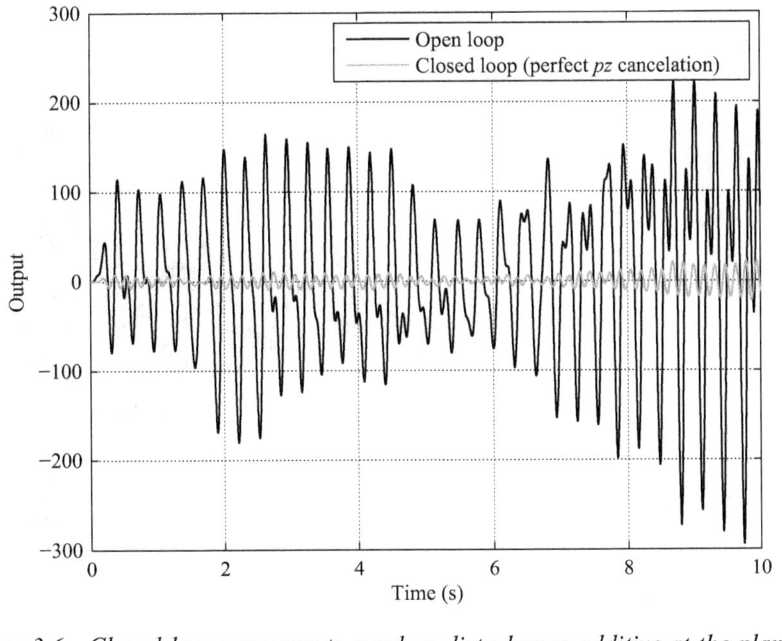

Figure 3.6 Closed-loop response to random disturbance additive at the plant input. Perfect pole/zero cancelation

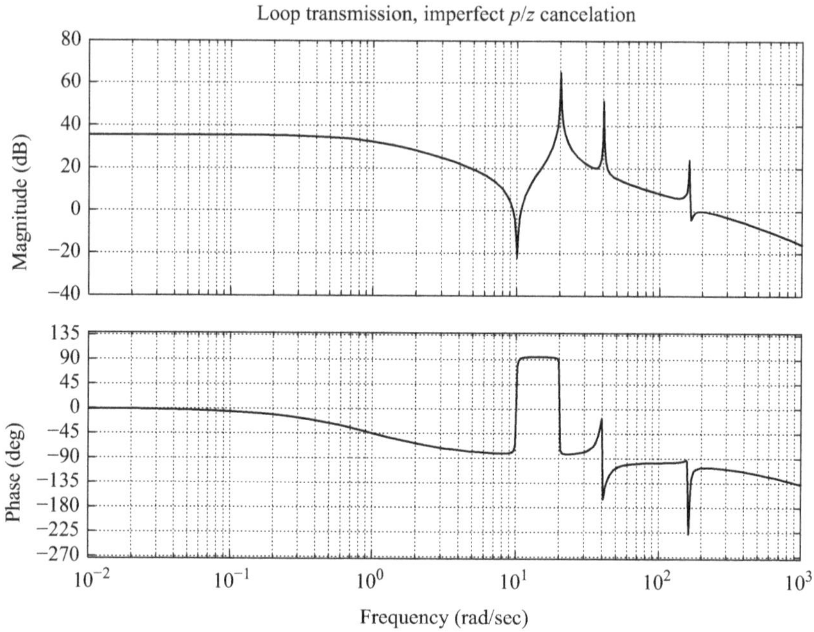

Figure 3.7 Loop transmission with imperfect pole/zero cancelation

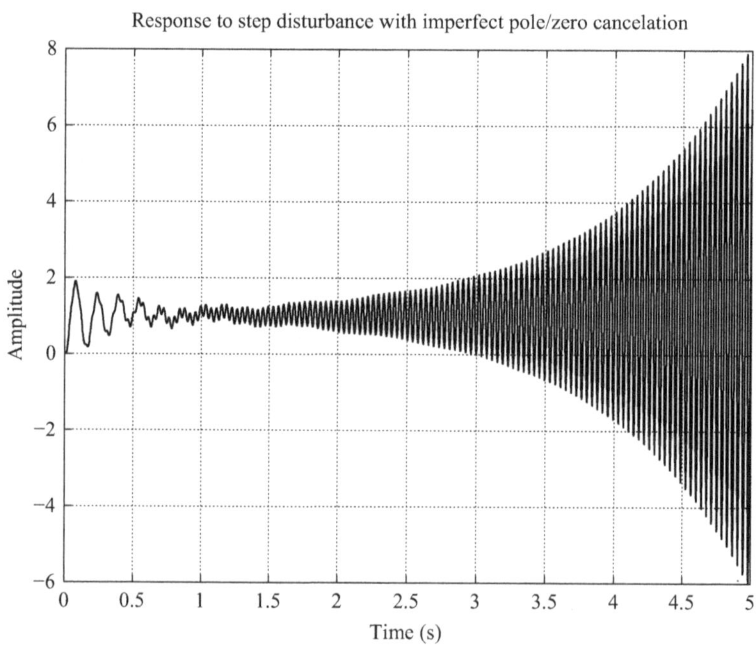

Figure 3.8 Closed loop response with imperfect pole/zero cancelation

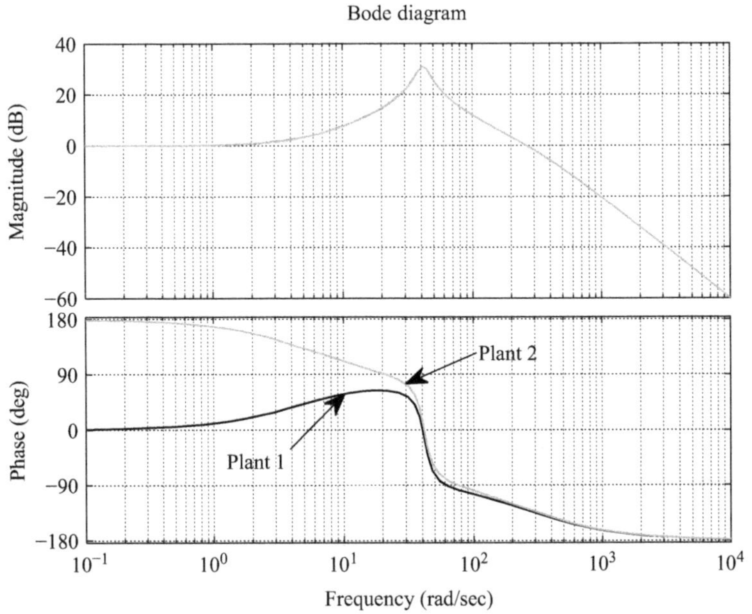

Figure 3.9 Frequency responses of two systems with identical moduli. The difference in argument is due to RHP zeros of the second system

at $s = -50$. The system with the zero in the RHP is *nonminimum phase*, as the system's phase lag is in excess of what is found using the Bode phase/gain relationship.

RHP zeros can have negative impact on feedback controller. While the two example systems have the same loop transmission moduli, Figure 3.10 shows that the sensitivity of the nonminimum phase system is very large at low frequency. This is due to the phase advance of the RHP zero at low frequency (180 degrees). It is noted that the phase advance of the RHP zero decays to 90 degrees at high frequency and is indistinguishable from a left half plane zero from this perspective.

3.4.5 Plant knowledge

Model-based control strategies are limited by the fidelity of the plant model. For example, a designer might have good knowledge (accurate modal frequency, quality factor, etc.) of the first few modes of a resonant plant, but limited information on subsequent modes. Clearly, this represents a bandwidth limitation, as the designer cannot assume that adequate margins of stability are achieved beyond the frequencies where there is accurate plant information. A model reduction strategy that ignores the dynamics at higher frequencies has the potential to cause control instability. For example, ignoring the last two modes from the second example in Section 3.4.3 would all but guarantee closed-loop instability if the bandwidth is high.

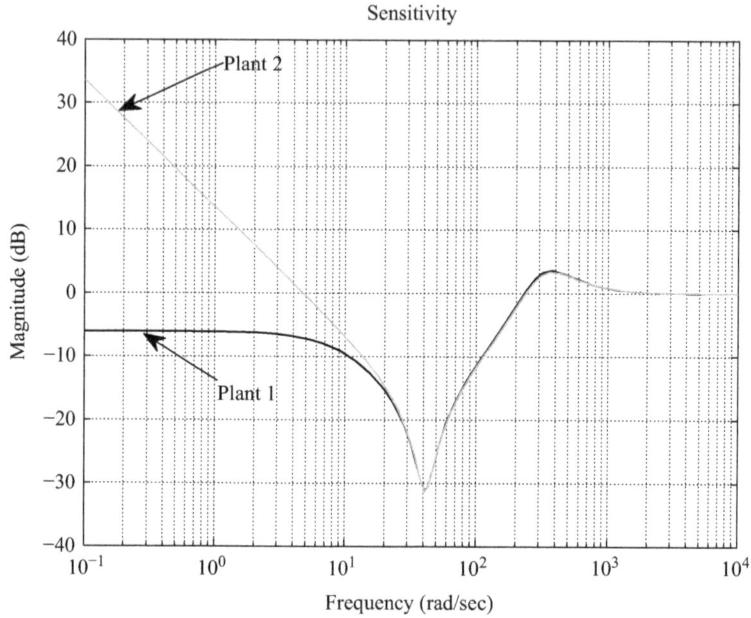

Figure 3.10 Sensitivity plot

3.4.6 *Time delay*

Transport lag or digital sampling are common causes for *time delay*, which is mathematically modeled by $G(s) = G_0(s)e^{-T_d s}$, where $G_0(s)$ is the transfer function of the nondelayed system, T_d is the time delay, and $G(s)$ is the frequency response of the delayed system. Clearly, $G(s)$ is nonminimum phase (inverse is not causal). The moduli of $G(s)$ and $G_0(s)$ are identical; however, the phase delay of $G(s)$ is greater than $G_0(s)$, the difference of which is a linear function of frequency. Figure 3.11 shows an experimentally acquired frequency response function of a system with time delay. Note the phase delay in excess of what is indicated by the Bode phase/gain relationship.

3.5 Exercises

1. State the advantages of large feedback.
2. What is the disturbance rejection of the open-loop controller?
3. The loop transmission at ω_1 is $T(j\omega_1) = -0.75 - j0.05$. The disturbance additive at the output is $10\ \sin(\omega_1 t)$. If there are no other exogenous inputs, what is the output of the feedback system as $t \to \infty$?
4. The loop transmission at ω_1 is $T(j\omega_1) = -60.5 - j0.1$. Is the feedback positive or negative?

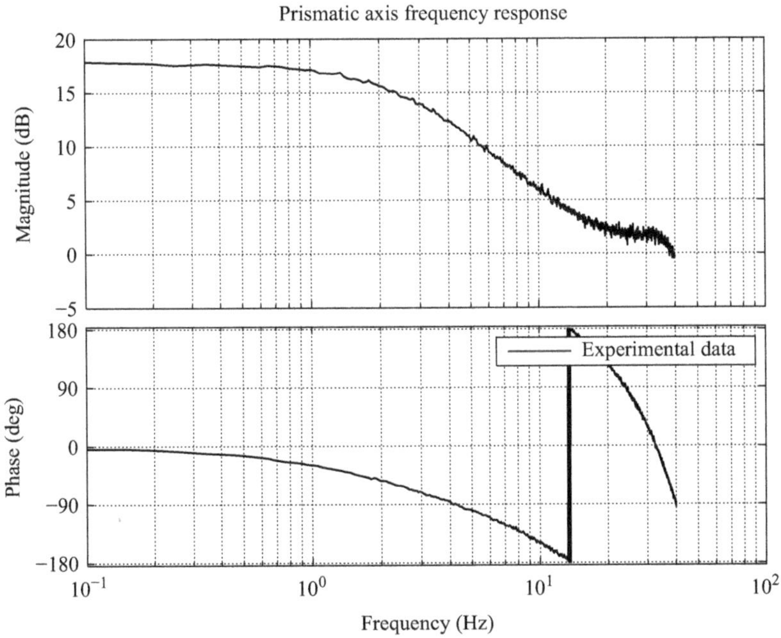

Figure 3.11 Experimentally acquired frequency response function

5. The loop transmission at ω_1 is $T(j\omega_1) = -125 + j0.01$. The plant gain at this frequency is uncertain to within $\pm 10\%$ of the nominal gain. What is the effect of this uncertainty on the tracking performance of the closed-loop system?
6. Explain how a large feedback system design can reduce system cost. Explain how it can increase the robustness of the system.
7. The loop transmission at ω_1 is $T(j\omega_1) = -0.9 - j0.01$. There is substantial disturbance power at ω_1. Qualify the closed-loop performance.
8. Explain the trade-off between expensive actuation and large feedback.
9. The loop transmission is $T(s) = \frac{1{,}000{,}000(s+1)}{(s+100)(s^2+s+100)}$. What is the sinusoidal steady-state response to a reference input $r(t) = 10\sin(2t)$? To $r(t) = 10\sin(2t)$?
10. For the system of the previous problem, find $\int_0^\infty \log|S(j\omega)|d\omega$.
11. The loop transmission of a feedback system is $T(s) = \frac{100}{s^2+s-1}$. Calculate $\int_0^\infty \log|S(j\omega)|d\omega$, where $S(j\omega)$ is the frequency response of the sensitivity function.
12. The loop transmission of a feedback system is $T(s) = \frac{100}{s^2+s-20}$. Calculate $\int_0^\infty \log|S(j\omega)|d\omega$, where $S(j\omega)$ is the frequency response of the sensitivity function.
13. Compare the closed-loop performance of the previous two feedback systems.
14. The sensor has monotone noise at 60 Hz of amplitude 2 V. The feedback at 60 Hz is 50 dB. Quantify the feedback system's sensor noise rejection.

15. Compare the sensitivity of $T_1(s) = 10^6 \frac{s+0.1}{(s+100)(s^2+s+100)}$ and $T_2(s) = 10^6 \frac{s-0.1}{(s+100)(s^2+s+100)}$.
16. The loop transmission phase delay for an analog control loop at frequency $\omega_1 = 600$ rad/s is -170 degrees. The analog compensator is replaced with a digital controller that introduces a time delay of 2 ms. What is the phase delay of the digital system at ω_1?

Chapter 4
Feedforward

I don't care if it oscillates on the step. It's *quiet*.

– *Anonymous*

It is clear from the material presented in the previous chapter that large feedback improves performance in several ways: better tracking, good disturbance rejection and reduced sensitivity to parameter variations in forward path systems. However, the Bode sensitivity integral indicates that for systems with greater than first-order roll-off (most real systems have this characteristic), the log integral of sensitivity over all frequencies is nonnegative, so negative feedback at some frequencies results in positive feedback at others. It is also understood that in practical applications, there are bandwidth limitations that in turn limit the feedback that can be applied.

In this chapter, methods that improve the performance of a feedback system by utilizing knowledge of system parameters are investigated. These are referred to as *feedforward* systems, although prefilters in series with a feedback system also belong to this category.

4.1 Command feedforward

Consider the tracking system in Figure 4.1. At frequencies where $|F| = |CAP| \gg 1$, the output y tracks the reference r very closely. However, outside these frequencies, the feedback system's performance degrades. Consider the case where the actuator and plant models $A_0(s)$ and $P_0(s)$ are accurate. This knowledge can be exploited by implementing a *command feedforward* system, shown in Figure 4.2.

The input–output function of this system is

$$\frac{y(s)}{r(s)} = \frac{FF(s)A(s)P(s) + C(s)A(s)P(s)}{1 + C(s)A(s)P(s)} \tag{4.1}$$

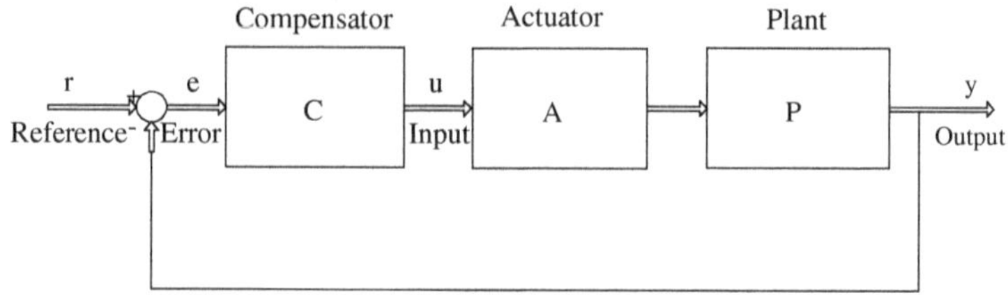

Figure 4.1 Block diagram of a tracking system

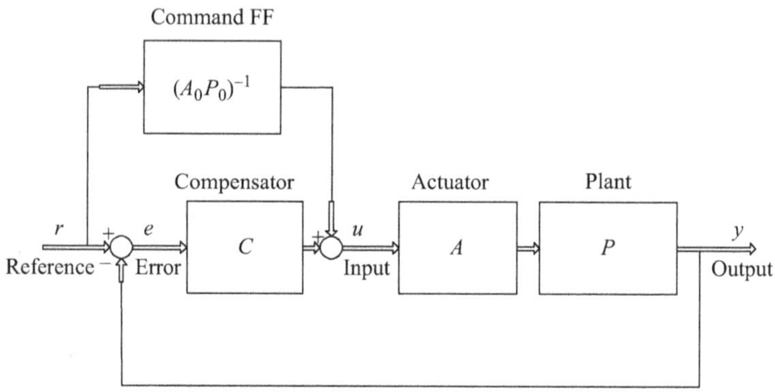

Figure 4.2 Block diagram of a command feedforward system

If $FF(s) = (A_0(s)P_0(s))^{-1}$, then

$$\frac{y(s)}{r(s)} = \frac{(A_0(s)P_0(s))^{-1}A(s)P(s) + C(s)A(s)P(s)}{1 + C(s)A(s)P(s)} \tag{4.2}$$

$$= \frac{1 + T_0(s)^{-1}}{1 + T(s)^{-1}} \tag{4.3}$$

where $T_0(s) = C(s)A_0(s)P_0(s)$. There is unity mapping of reference to output in two conditions:

1. $|C(s)| \to \infty$. In this case, the feedback is large, so both $T_0^{-1}(s)$ and $T^{-1}(s)$ vanish.
2. The actuator and plant models, $A_0(s)$ and $P_0(s)$, do not deviate substantially from $A(s)$ and $P(s)$, and $T_0\ (s) \simeq T(s)$. This results in good tracking at frequencies where feedback is not necessarily large.

The command feedforward system improves tracking performance at frequencies where feedback cannot be large. However, careful consideration of what exactly the feedback bandwidth limitations are before implementing command feedforward is indicated. Recall these limiting characteristics from Chapter 3.

1. Actuator limits. If, for example, the actuator is rate limited, command feedforward may result in excessive saturation.
2. Plant limits – zeros. If the plant has RHP zeros, then $P_0^{-1}(s)$ will be an unstable system.
3. Plant knowledge. Although an accurate model of the actuator may be easy to develop, the plant model might not be. At frequencies where $P_0(s)$ is not close to $P(s)$, the advantage of command feedforward is lost.
4. Time delay. Time delay, T_d, in the plant (e.g. the sensor is digital) requires the feedforward system to include $e^{T_d s}$, which violates the condition of causality. It is possible to ameliorate the effect of time delay at low frequency compared to the sample frequency; however, the exponential function describing the delay clearly cannot be included in the system $FF(s)$.

In addition to these feedback limitations as they relate to command feedforward, the system $FF(s)$ should be strictly proper. In most cases, both the actuator and plant frequency responses decrease at high frequency, so the inverse of the transfer functions that model these responses would not be proper. This can be mitigated by including poles in $FF(s)$ at frequencies much higher than the operational frequency of the feedforward/feedback system.

It is evident that command feedforward is practical in only a few cases, primarily those that involve bandwidth limitations driven by sensor inadequacies.

Example: Command feedforward
Consider a single-input, single-output (SISO) LTI plant $P(s) = \frac{20{,}000}{s(s^2+s+100)}$. Sensor limitations limit the bandwidth to 80 rad/s. A sixth-order feedback compensator $C(s)$ is designed to crossover at 85 rad/s, provide more than 60 dB of feedback at 1 rad/s and roll off at fourth order at high frequency. The loop transmission $T(s) = C(s)P(s)$ is shown in Figure 4.3.

Figure 4.4 shows the response to a unit amplitude, 5 rad/s sinusoidal input. There is nearly 40 dB of feedback at this frequency, and thus the system tracks the input very accurately. Figure 4.5 shows the response of the same system to the unit amplitude, 50 rad/s sinusoidal input. The loop gain at this frequency is less than 2, and thus the tracking lacks accuracy. As the sensor provides a limit on bandwidth, the tracking performance at this frequency cannot be improved by increasing feedback.

A command feedforward system is designed to improve the tracking performance at frequencies close to crossover. The actuator is assumed to be 1, so the feedforward filter is simply the inverse of the plant transfer function. However, $P(s)^{-1}$ is not proper. So additional poles are applied at frequencies well beyond the crossover to make this system strictly proper without affecting the phase significantly near crossover, $FF(s) = P(s)^{-1}\frac{10{,}000^3}{(s+10{,}000)^3}$. Figure 4.6 shows the response to the 50 rad/s reference with the command feedforward system in place. The tracking performance is greatly improved at this frequency in the close neighborhood of the feedback system crossover.

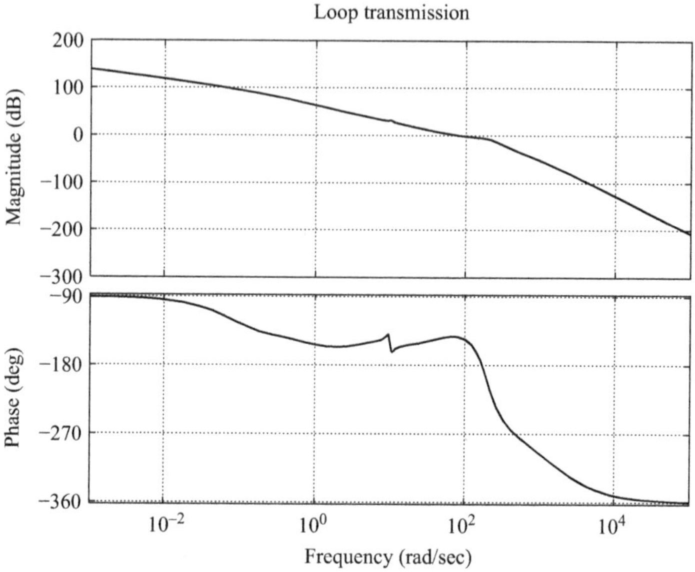

Figure 4.3 Loop transmission of the feedback system

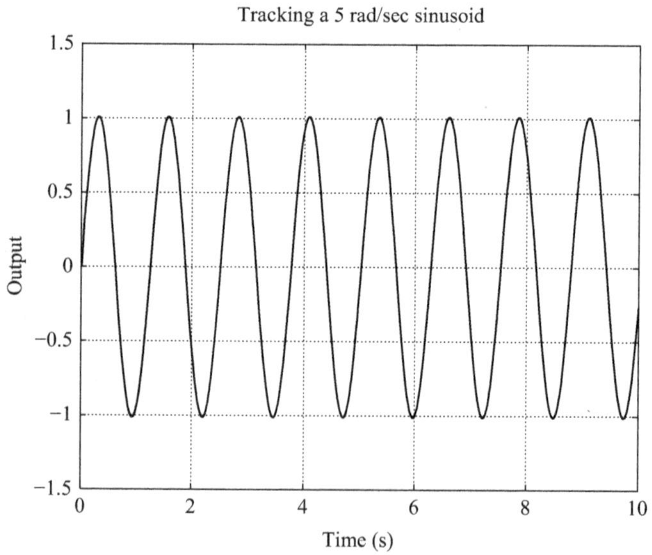

Figure 4.4 Response of feedback system to a 5 rad/s sinusoid (amplitude 1)

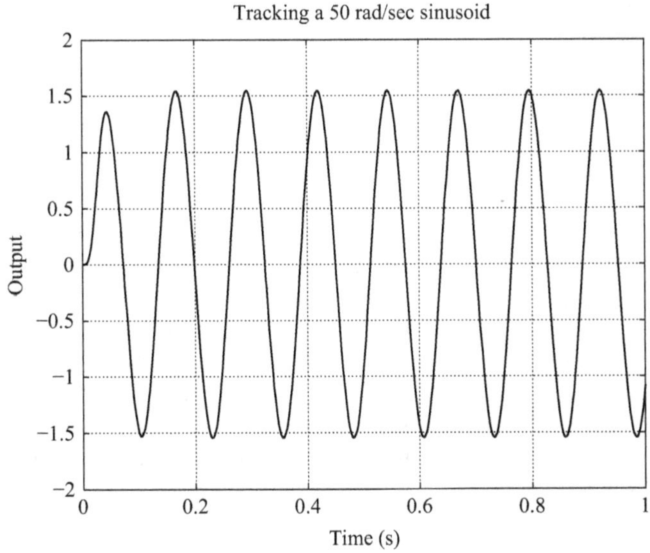

Figure 4.5 Response of feedback system to a 50 rad/s sinusoid (amplitude 1)

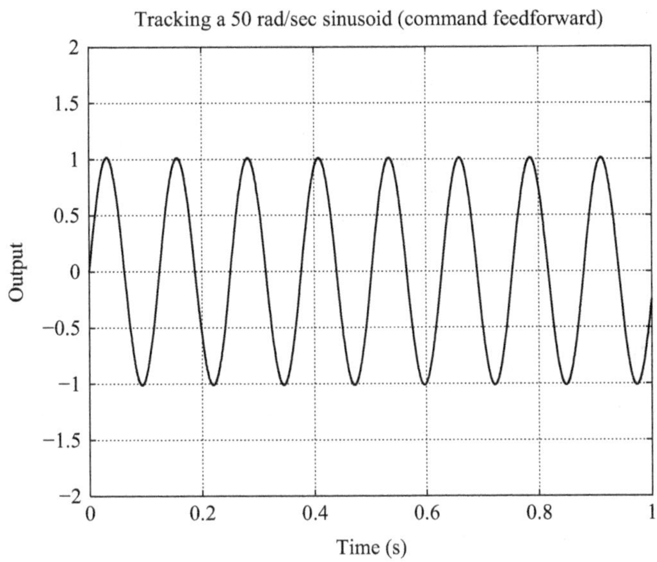

Figure 4.6 Response of feedback/command feedforward system to a 50 rad/s sinusoid (amplitude 1)

4.2 Prefilter

The loop transmission function for a tracking system is as follows:

$$T(s) = (0.125 \times 10^7)\frac{s+800}{(s+2000)(s^2+10s+100)} \tag{4.4}$$

The frequency response of $T(s)$ is shown in Figure 4.7. The bandwidth is $\simeq$ 820 rad/s, with good margins of stability. Sinusoidal disturbances in the functional bandwidth of 10 rad/s are attenuated by a factor of 5000, which is excellent performance. However, the response of the closed-loop system to the reference step has substantial overshoot, as seen in Figure 4.8. The question arises: How should the control designer modify the feedback compensator to improve the transient response? Clearly, the overshoot can be reduced if the loop gain is lowered, indicating a possible trade-off between disturbance rejection and transient response performance.

The correct answer to this question is ... *nothing*. The complete design is a sequence of subdesigns. First, the feedback compensator $C(s)$ is designed to maximize available feedback; second, a prefilter $R(s)$ is designed to modify the closed-loop response of the system. Figure 4.9 is a block diagram of a feedback system with prefiltering. The closed-loop response of the system is

$$\frac{y(s)}{r(s)} = R(s)\frac{T(s)}{T(s)+1} \tag{4.5}$$

A common prefilter is the *notch*.

$$R(s) = \frac{s^2 + 2\pi f_b s + (0.9(2\pi f_b))^2}{s^2 + 2(2\pi f_b)s + (0.9(2\pi f_b))^2} \tag{4.6}$$

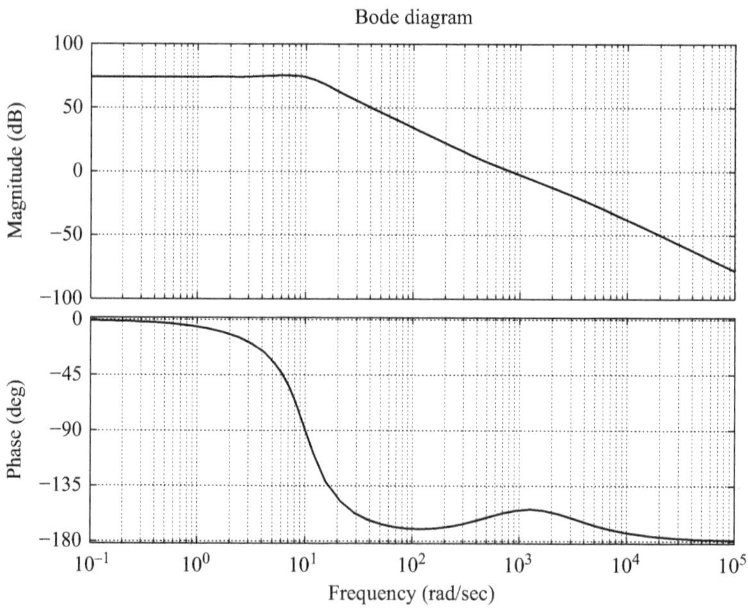

Figure 4.7 Frequency response

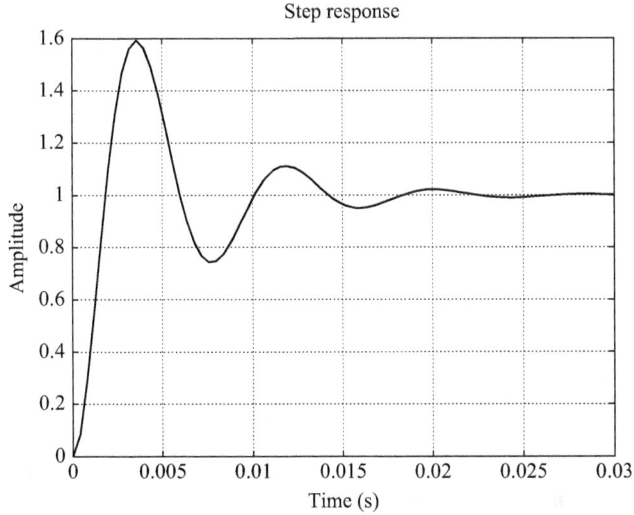

Figure 4.8 Step response w/o prefilter

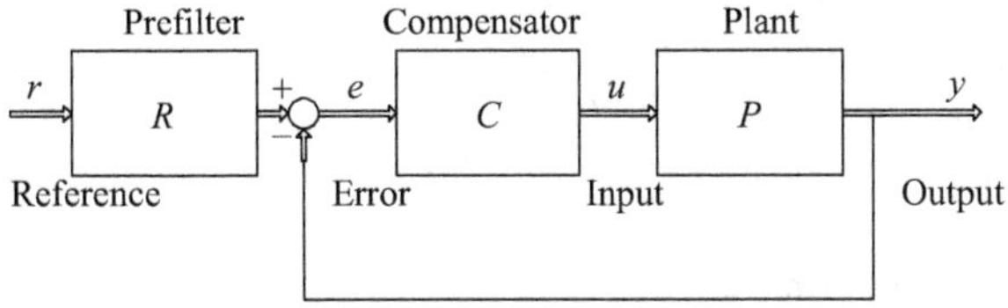

Figure 4.9 Block diagram of a feedback control system with a prefilter

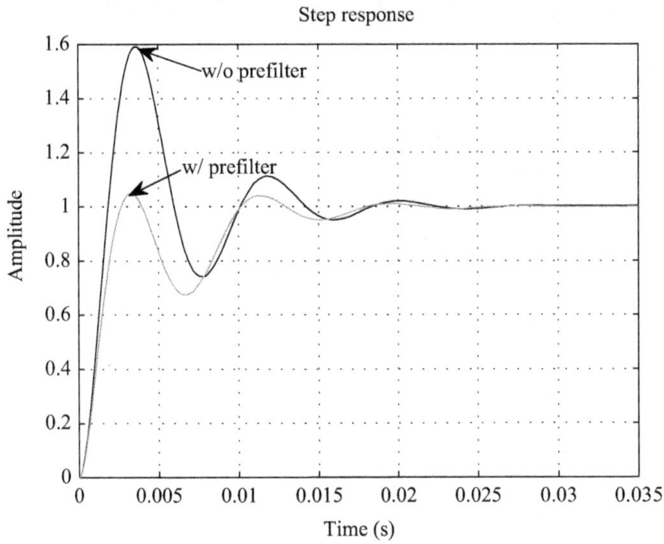

Figure 4.10 Step response with a prefilter

where f_b is the 0 dB loop transmission crossover frequency (in Hz). Figure 4.10 shows the step reference response of the example tracking system with the notch prefilter. The overshoot is substantially reduced without reducing the feedback (no reduction in disturbance rejection over the functional bandwidth).

It is noted that command feedforward and prefiltering are mathematically equivalent if $R(s) = C^{-1}(s)FF(s) + 1$. However, improving the reference tracking performance requires a lower-order system with command feedforward if the feedback compensator is strictly proper (its inverse is not proper, requiring more poles in $R(s)$ at high frequency for strictly proper filter).

4.3 Exercises

1. The feedback bandwidth is 200 Hz. How good is the tracking performance at 180 Hz? Why?
2. A tracking system must be accurate to 30 Hz. Because of bandwidth restrictions, the feedback at 30 Hz is 6 dB. Comment on the performance of this tracking system.
3. Describe the disturbance rejection improvement achieved by the inclusion of command feedforward to a tracking system.
4. List three characteristics of a feedback tracking system that make it a poor candidate for command feedforward.
5. The plant has an unstable pole. Describe the transient response characteristics of the command feedforward system. Focus on the deleterious effects of the RHP zero.
6. The loop transmission function is

$$T(s) = \frac{10^9}{s^5+211s^4+1.221\times 10^4 s^3+2.112\times 10^6 s^2+2.21\times 10^7 s+2\times 10^7}.$$

   Use Matlab to plot the frequency response of $T(s)$. Quantify the tracking performance at 1, 20, 100, and 150 rad/s. Explain the contrast in performance at these frequencies.
7. For the system of Problem 6, write a Matlab simulation to find the tracking system output to sinusoidal inputs at 1, 20, 100, and 150 rad/s.
8. The system with loop transmission $T(s)$ in Problem 6 has insufficient tracking performance at 20 rad/s. Because of feedback limitations and robustness requirements, the feedback cannot be increased. Design a command feedforward system to improve the performance. The feedforward system must be strictly proper. Use Matlab to plot the response to a sinusoidal input at 20 rad/s. Compare this to the response of the original feedback controller.
9. For the feedback tracking system of Problem 6, use Matlab to plot the step response of the closed-loop system. What is the overshoot? Suggest a design change that does not reduce the feedback of the tracking system.
10. Design a prefilter for the tracking system of Problem 6 to improve the overshoot. Use Matlab to compare the transient performance of the feedback system and the feedback system with the prefilter.

Chapter 5
Stability

If the plant is stable, and the retention of stability is the only concern, then leave the system open loop.

– *Anonymous*

It is stated in Chapter 3 that feedback improves disturbance rejection and tracking and reduces sensitivity to parameter variations. In this chapter, the effect of feedback on system stability is developed. First, definitions of stability for linear systems must be developed both in the frequency and time domains. The Nyquist Stability Criterion, an extremely powerful analysis tool in determining not only the stability of a feedback system but also its relative stability, is presented. The generalized Nyquist Criterion is also discussed along with Gershgorin's Theorem for the stability analysis of multivariable systems. The deleterious effect of hidden unstable pole/zero cancelations that Nyquist analysis cannot detect is addressed in defining the important concept of *internal* stability. The Lyapunov method for stability analysis of nonlinear systems is presented so that a treatment of *absolute* stability can be included. The latter is of critical importance in the development of nonlinear compensators discussed in Chapter 7.

5.1 Bounded-input, bounded-output stability

A linear, time invariant (LTI) system with zero initial conditions is said to be bounded-output, bounded-input (BIBO) stable if every bounded input causes a bounded output. Although the definition is useful in providing a negative result, it is not terribly useful in determining if a system is stable.

A single-input, single-output (SISO), LTI, causal and zero-state system is BIBO stable if and only if its impulse response, $g(t)$, is absolutely integrable on the interval $[0, \infty)$.

Sufficiency

The integral of the absolute value of the impulse response is bounded.

$$y(t) = \int_0^{\infty} |g(\tau)| d\tau \leq N < \infty \tag{5.1}$$

The zero-state response, $y(t)$, to input $u(t)$ is found by the following convolution integral:

$$y(t) = \int_0^\infty u(t-\tau)g(\tau)d\tau \tag{5.2}$$

Take the absolute value of the response.

$$|y(t)| = \left|\int_0^\infty u(t-\tau)g(\tau)d\tau\right| \leq \int_0^\infty |u(t-\tau)||g(\tau)|d\tau \tag{5.3}$$

The input is bounded, thus for some $M < \infty$, $|u(t)| < M$, $\forall t$.

$$|y(t)| \leq M\int_0^\infty |g(\tau)|d\tau \leq MN < \infty \tag{5.4}$$

The last inequality is a consequence of $g(t)$ being absolutely integrable.

Necessity
Assume the system is BIBO stable, but $g(t)$ is not absolutely integrable. For any arbitrarily large N, there exists some upper limit of integration, t_1, such that $\int_0^{t_1} |g(\tau)|d\tau \geq N$. Select a bounded input $u(t_1 - \tau) = sgn(g(\tau))$, where sgn is the signum function. The output is arbitrarily large, $y(t_1) = \int_0^\infty g(\tau)u(t_1-\tau)d\tau = \int_0^\infty |g(\tau)|d\tau > N$ This contradicts the assumption of BIBO stability.

Stability of SISO, LTI systems is determined finding the roots of the characteristic polynomial. A SISO, LTI system with a proper rational transfer function $G(s)$ is BIBO stable if and only if every pole of $G(s)$ has a negative real part (open left half plane (OLHP)). If $G(s)$ has a pole, p_i, with multiplicity, m_i, the partial fraction expansion has the factors $\frac{1}{s-p_i}, \frac{1}{(s-p_i)^2} \cdots \frac{1}{(s-p_i)^{m_i}}$. The inverse Laplace transforms of these terms are $e^{p_i t}, te^{p_i t} \ldots t^{m_i-1}e^{p_i t}$. These are absolutely integrable if and only if p_i has a negative real part.

5.1.1 Marginally stable systems

Linear systems with distinct poles with zero real parts (on the $j\omega$ axis) and all other poles having negative real parts are referred to as *marginally* stable. These systems have bounded responses to most bounded inputs; however, particular bounded inputs will result in unbounded outputs. Systems with an integrator (a single pole at the origin with all other poles in the OLHP) are marginally stable. Consider a linear system with the following transfer function:

$$G(s) = \frac{10}{s(s+10)} \tag{5.5}$$

The poles are at $s = 0$ and $s = -10$. The Laplace transform of the response of this system to the unit step function is

$$y(s) = \frac{10}{s^2(s+10)} \tag{5.6}$$

The partial fraction expansion of $y(s)$ is

$$y(s) = \frac{-0.1}{s} + \frac{1}{s^2} + \frac{0.1}{s+10} \tag{5.7}$$

The inverse Laplace transform of $y(s)$ is

$$y(t) = \left(-0.1 + t + 0.1e^{-10t}\right)1^+ \tag{5.8}$$

where 1^+ is the unit step function. The second term is the ramp function, which is unbounded. So, the marginally stable system is not BIBO stable. However, the response of this system to other bounded inputs is bounded. For instance, the response of the same system to the bounded exponential input e^{-5t} is

$$y(t) = \left(0.2 - 0.4e^{-5t} + 0.2e^{-10t}\right)1^+ \tag{5.9}$$

which is bounded. It is noted that the marginally stable systems differ from systems with right half plane (RHP) poles. For example, the response of the linear system with transfer function $G(s) = \frac{10}{s-10}$ to a bounded input will have an unbounded exponential term e^{10t} (the natural response of the system is always divergent). Similarly, marginally stable systems differ from systems with left half plane (LHP) poles and repeated poles on the $j\omega$ axis (e.g. the response to the linear system with transfer function $G(s) = \frac{10}{s^2(s+10)}$ to bounded inputs will include the ramp function as part of the natural response).

5.1.2 *BIBO stability of state equations*

The definition of BIBO stability is expanded to the time domain description of LTI systems. Given the (A, B, C, D) state space realization, the transfer matrix is found.

$$\begin{aligned} \dot{x}(t) &= Ax(t) + Bu(t) \\ y(t) &= Cx(t) + Du(t) \end{aligned}$$

Take the Laplace transform of the state differential equation to find the response.

$$\begin{aligned} sIX(s) &= AX(s) + BU(s) \\ X(s) &= (sI - A)^{-1}BU(s) \\ CX(s) &= C(sI - A)^{-1}BU(s) \\ Y(s) &= C(sI - A)^{-1}BU(s) + DU(s) \end{aligned}$$

The transfer matrix mapping $U(s)$ to $Y(s)$ is

$$G(s) = C(sI - A)^{-1}B + D \tag{5.10}$$

$$= \frac{1}{det(sI - A)} Cadj(sI - A)B + D \tag{5.11}$$

where *adj*(·) is the adjoint matrix. The multiple-input, multiple-output (MIMO) system with transfer matrix is BIBO stable if and only if the poles of each rational function element all have negative real parts.

It is evident that the poles of transfer matrix $G(s)$ are eigenvalues of the state matrix A. However, eigenvalues of A are not necessarily poles of $G(s)$, as there is the possibility of common factors of elements of $Cadj(sI - A)B$ and the polynomial $det(sI - A)$, which is the case for state space realizations that are not minimal. Thus, even if A has eigenvalues with positive real parts, the system is BIBO stable if there is cancelation of the unstable factors of $det(sI - A)$.

5.2 Zero input stability

Consider the LTI system with zero input and nonzero initial state x_0. The solution of the state differential equation $\dot{x} = Ax(t)$ is $x(t) = e^{At}x_0$.

Definition: Marginal stability
The zero input response of $\dot{x} = Ax$ is marginally stable if every finite initial state x_0 excites a bounded response.

Definition: Asymptotic stability
The zero input response is asymptotically stable if every finite initial state excites a bounded response that approaches zero as $t \to \infty$.

System $\dot{x} = Ax$ is asymptotically stable if and only if all eigenvalues of A have negative real parts. It is evident that asymptotic stability implies BIBO stability; however, the converse is not necessarily true. It is marginally stable if and only if all eigenvalues of A have zero or negative real parts, and those with zero real parts are simple roots of the minimal polynomial of A. The method using the Lyapunov Theorem to determine asymptotic stability has been explained in Reference 3.

5.2.1 Hidden modes

Consider a system consisting of a cascade of two SISO systems, figure 5.1. The unstable pole of the second system is canceled by the nonminimum phase zero of the first system, and the transfer function of the resulting BIBO stable system is $G(s) = \frac{y(s)}{u(s)} = \frac{1}{s+1}$. We now find two different state space realizations:

$$X(s) = \frac{1}{s+1}U(s) \tag{5.12}$$

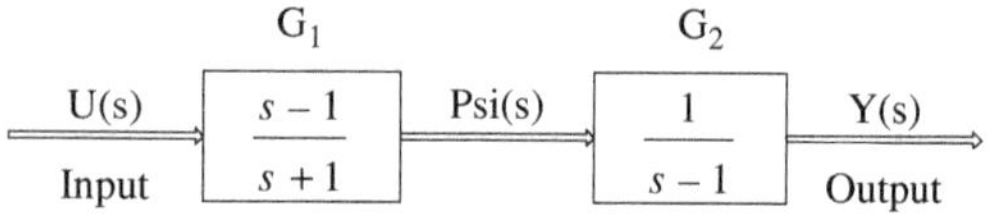

Figure 5.1 Linear system with pole/zero cancellation

Employing the derivative property of the Laplace transform yields the following differential equation:

$$\dot{x}(t) + x(t) = u(t) \tag{5.13}$$
$$\dot{x}(t) = -x(t) + u(t) \tag{5.14}$$

Now we solve for the intermediate signal ψ.

$$\Psi(s) = (s - 1)X(s) \tag{5.15}$$
$$\psi(t) = \dot{x}(t) - x(t) \tag{5.16}$$
$$= -x(t) + u(t) - x(t) \tag{5.17}$$
$$= -2x(t) + u(t) \tag{5.18}$$

Now we find the derivative of the output $y(t)$.

$$Y(s) = \frac{1}{(s-1)}\Psi(s) \tag{5.19}$$
$$\dot{y}(t) - y(t) = -2x(t) + u(t) \tag{5.20}$$
$$\dot{y}(t) = -2x(t) + y(t) + u(t) \tag{5.21}$$

Now the state equations can be expressed.

$$\begin{bmatrix} \dot{x} \\ \dot{y} \end{bmatrix} = \begin{bmatrix} -1 & 0 \\ -2 & 1 \end{bmatrix}\begin{bmatrix} x \\ y \end{bmatrix} + \begin{bmatrix} 1 \\ 1 \end{bmatrix}u, \quad y = \begin{bmatrix} 0 & 1 \end{bmatrix}\begin{bmatrix} x \\ y \end{bmatrix} \tag{5.22}$$

The observability matrix $\begin{bmatrix} 0 & 1 \\ -2 & 1 \end{bmatrix}$ is rank 2, whereas the controllability matrix $\begin{bmatrix} 1 & -1 \\ 1 & -1 \end{bmatrix}$ is rank 1. Thus, the state space realization is not minimal. In this case, the unstable block is separated from the input by the nonminimum phase zero in the second block.

Consider the same system (from the perspective of input–output transfer function), but with the order of the blocks reversed.

$$\Psi(s) = \frac{1}{s-1}U(s) \tag{5.23}$$
$$\dot{\psi}(t) - \psi(t) = u(t) \tag{5.24}$$
$$y(s) = \frac{s-1}{s+1}\psi(s) \tag{5.25}$$

$$\dot{y}(t) + y(t) = \dot{\psi}(t) - \psi(t) \tag{5.26}$$
$$= u(t) \tag{5.27}$$
$$\dot{y}(t) = -y(t) + u(t) \tag{5.28}$$

The state equations for this system are

$$\begin{bmatrix} \dot{\psi} \\ \dot{y} \end{bmatrix} = \begin{bmatrix} 1 & 0 \\ 0 & -1 \end{bmatrix} \begin{bmatrix} \psi \\ y \end{bmatrix} + \begin{bmatrix} 1 \\ 1 \end{bmatrix} u, \quad y = \begin{bmatrix} 0 & 1 \end{bmatrix} \begin{bmatrix} \psi \\ y \end{bmatrix}$$

The controllability matrix $\begin{bmatrix} 1 & 1 \\ 1 & -1 \end{bmatrix}$ is rank 2, whereas the observability matrix is $\begin{bmatrix} 0 & 1 \\ 0 & -1 \end{bmatrix}$ is rank 1. This state space realization is not minimal. In that case, the unstable pole is separated from the output by the nonminimum phase zero.

Systems that are BIBO stable with hidden unstable poles have a particularly sinister quality. For the second system, consider the internal signal $\psi(t)$ with a nonzero, finite initial value and $u(t) = 0$.

$$\dot{\psi}(t) = \psi(t) \tag{5.29}$$

$$\psi(t) = e^{t}\psi(0) \tag{5.30}$$

The response to this initial condition is unbounded. Although stable from the input–output perspective, systems with unstable pole/zero cancelations lack zero input stability and feedback systems will lack *internal* stability, to be discussed later. As such, unstable poles should never be canceled by introducing zeros at the same location in the complex plane.

5.3 Nyquist Stability Criterion

The stability of feedback systems is now investigated. The location of the zeros of $F(s) = 1 + T(s)$, which are the location of the closed-loop poles, can be determined in a number of ways. The Routh Criterion can be used to determine if there are RHP poles. The root locus method finds the location of closed-loop poles as a function of a variable parameter. Although effective, these methods provide little in the way of insight into the relationship of the loop transmission function and the condition of closed-loop stability.

Cauchy's Theorem (Principle of the Argument)

If a closed contour Γ_s in the s-plane encircles Z zeros and P poles of $F(s)$ and does not pass through any poles or zeros of $F(s)$ and the traversal about the contour is clockwise, the corresponding contour Γ_F in the $F(s)$ plane encircles the origin of this plane $N = Z - P$ times in the clockwise direction.

The mapping of a closed contour through function $F(s)$ reveals the relative number of zeros and poles of the function residing within the contour without direct calculation of polynomial roots. Consider a contour in the s-plane beginning at the origin, proceeding up the positive imaginary axis to $j\infty$, sweeping clockwise in a semicircle of infinite radius to $-j\infty$, then back to the origin along the negative imaginary axis. This contour, Γ_N, encloses the entire RHP and is called the *Nyquist contour*. If $F(s)$ is the return difference of a feedback system, Cauchy's Theorem can be used to determine closed-loop stability. Note that the zeros of $F(s)$ in this case are the closed-loop poles of the feedback system. For BIBO stability, it is a requirement that no zeros of $F(s)$ reside in the RHP.

Nyquist Stability Criterion
A feedback control system is stable if and only if for the contour Γ_F (called the Nyquist plot) found by mapping the Nyquist contour Γ_N through the return ratio $F(s)$, the number of anticlockwise encirclements of the origin of the F-plane is equal to the number of poles of $F(s)$ in the open right half plane (ORHP).

This follows from Cauchy's Theorem for the specific case of mapping the Nyquist contour through the return difference. For stability, the number of zeros of $F(s)$ within the Nyquist contour must be zero (i.e. $Z = 0$). So, the number of F-plane origin encirclements must be $N = 0 - P = -P$. Note that P is the number of poles of the return difference within the Nyquist contour. Recall that the return difference is the loop transmission plus 1, (i.e. $1 + T(s)$). If $T(s) = \frac{n(s)}{d(s)}$, then $F(s) = \frac{n(s)+d(s)}{d(s)}$, and the poles of the return difference are the poles of the loop transmission. So, only the number of open-loop RHP poles (P) need be known to assess the stability of the closed-loop system. The F-plane plot must encircle the origin P times in the anticlockwise direction.

It is evident that the Nyquist Stability Criterion may be implemented when the Nyquist contour is mapped through the loop transmission function $T(s)$. As $T(s) = F(s) - 1$, the T-plane plot must encircle the point $-1 + j0$ P times anticlockwise as a necessary and sufficient condition for closed-loop stability. This is typically the plot that is utilized, as the loop transmission function is more easily calculated. In addition, for some systems where a model is not readily available, the frequency response of the loop opened at convenient point may be experimentally measured. Note that the polar plot of this response represents a part of the Nyquist plot (the mapping of the positive $j\omega$ axis through the loop transmission). For strictly proper systems, this represents half the plot. The missing half (the mapping of the negative imaginary axis) is symmetric about the real axis of the T-plane plot. As such, the point -1 in the T-plane is particularly significant, warranting the name *critical point*.

The effect of a scalar gain, k, applied to the loop transmission function is easily determined by analyzing the T-plane plot. $k > 1$ expands the plot away from the origin by this factor at all points, $0 < k < 1$ shrinks the plot by this factor. A glance at the T-plane plot ($k = 1$) shows how much gain increase or decrease is possible without encircling the critical point. For $k < 0$, the T-plane plot flips about the imaginary axis, retaining the original contour direction. For example, consider $T(s) = \frac{1}{s+5}$. The frequency response is

$$T(j\omega) = \frac{1}{j\omega + 5} \tag{5.31}$$

$$= \frac{5 - j\omega}{\omega^2 + 25} \tag{5.32}$$

The plot of this is half the Nyquist plot, shown in Figure 5.2. Now consider $kT(s)$, $k = -1$. The frequency response is

$$T(j\omega) = \frac{-1}{j\omega + 5} \tag{5.33}$$

$$= \frac{-5 + j\omega}{\omega^2 + 25} \tag{5.34}$$

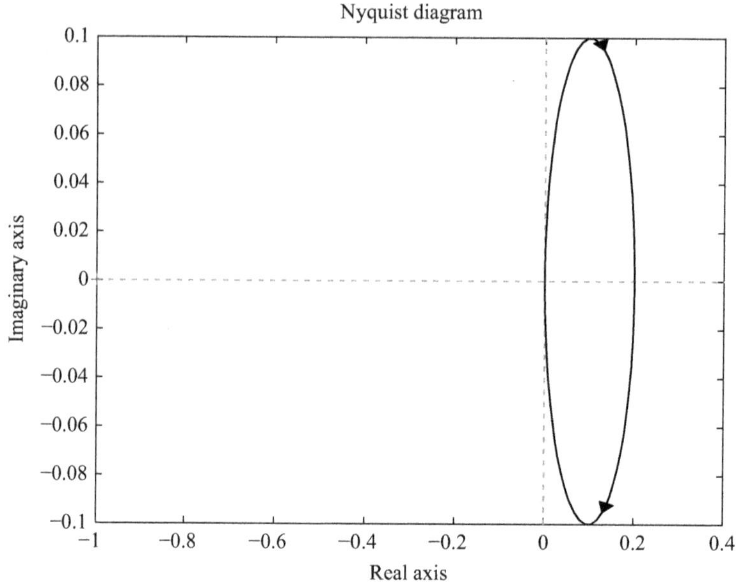

Figure 5.2 Nyquist plot of $\frac{1}{s+5}$

The negation moves the points to the diagonal quadrant, where the graph proceeds in the same direction as the original plot. This is the effect of adding 180 degrees of phase delay (sign inversion) to the original system *T*(*s*) (Figure 5.3).

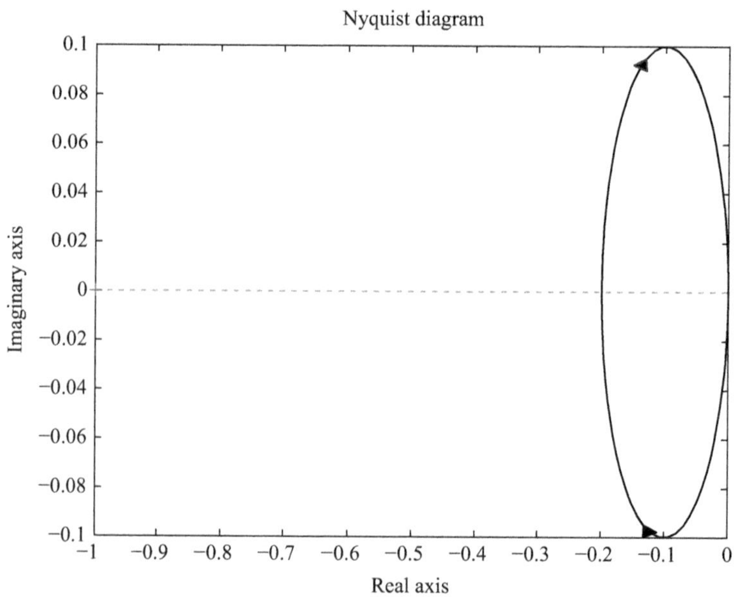

Figure 5.3 Nyquist plot of $(-1)\,\frac{1}{s+5}$

Examples: Nyquist Stability Criterion

Figure 5.4 shows the T-plane plot of $T(s) = \frac{100}{(s+1)(s+10)}$. The loop transmission function is stable, thus the closed-loop system is stable by the Nyquist Criterion if there are no net encirclements of the critical point. The phase of this function approaches -180 degrees at high frequency, so the critical point cannot be encircled regardless of whatever gain increase is applied. The closed-loop system is stable.

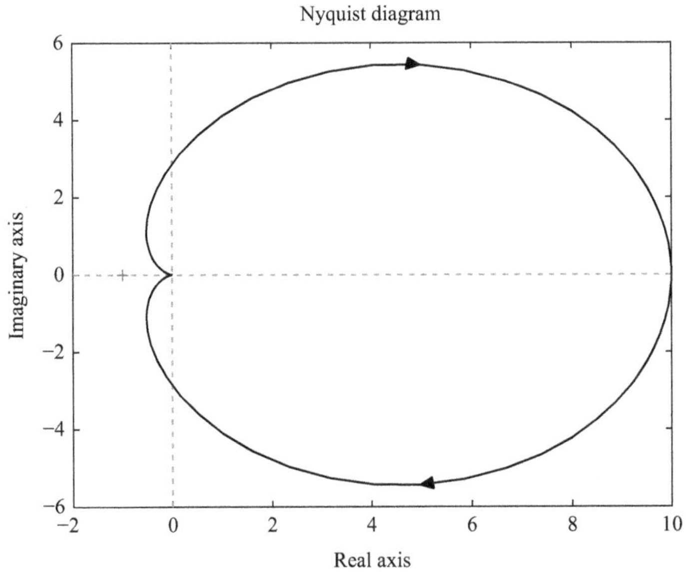

Figure 5.4 Nyquist plot of $\frac{100}{(s+1)(s+10)}$

Figure 5.5 is the T-plane plot of the loop transmission function $T(s) = \frac{0.5}{s-1}$. There is one RHP pole of this system, thus by the Nyquist Criterion the T-plane plot must encircle the critical point one time in the anticlockwise direction (it does not). It is evident that if the gain is increased by $k > 2$, the T-plane plot encircles -1 one time counterclockwise, and the system is stable.

Consider the loop transmission function $T(s) = \frac{100}{(s-1)(s-10)}$. The system has two RHP poles, and thus the Nyquist locus must encircle the critical point twice in the counterclockwise direction. Figure 5.6 shows the Nyquist plot of this function. It is evident that the system is unstable. In addition, an increase in gain does not stabilize the system as it did in the prequel, as the Nyquist locus does not cross the negative real axis and thus cannot encircle the critical point. If a sign inversion is introduced into the loop, the Nyquist locus rotates about the imaginary axis and retains its direction. The plot now encircles the critical point one time in the anticlockwise direction. The closed-loop system is still unstable, with a single unstable pole as opposed to two.

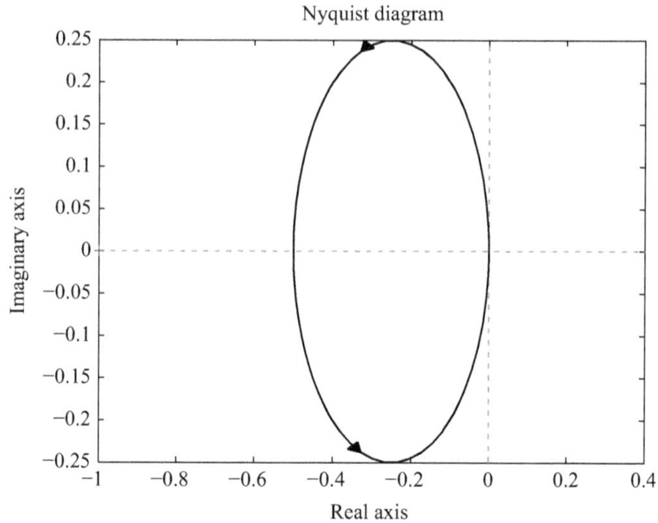

Figure 5.5 Nyquist plot of $\frac{0.5}{s-1}$

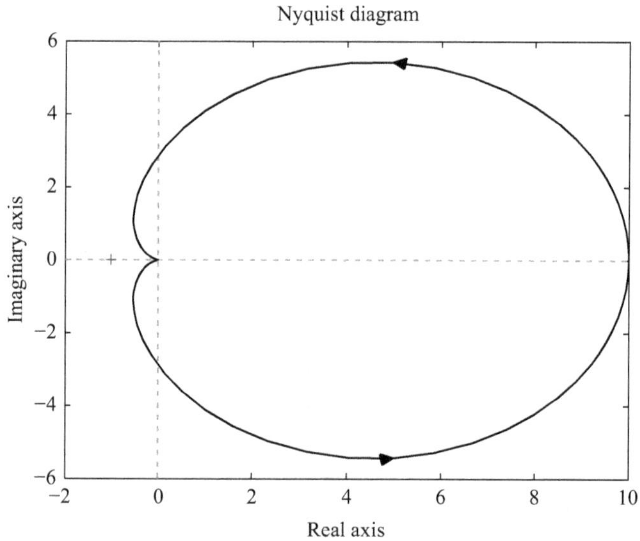

Figure 5.6 Nyquist plot of $\frac{100}{(s-1)(s-10)}$

Figure 5.7 shows the Nyquist plot of loop transmission function $T(s) = \frac{100(s-1)}{s^2+10s+100}$. System $T(s)$ is stable, so zero net encirclements of the critical point is necessary for closed-loop stability. Note the effect of the RHP zero in the loop transmission. The DC gain and phase are 1 and 180°, respectively, and the Nyquist locus

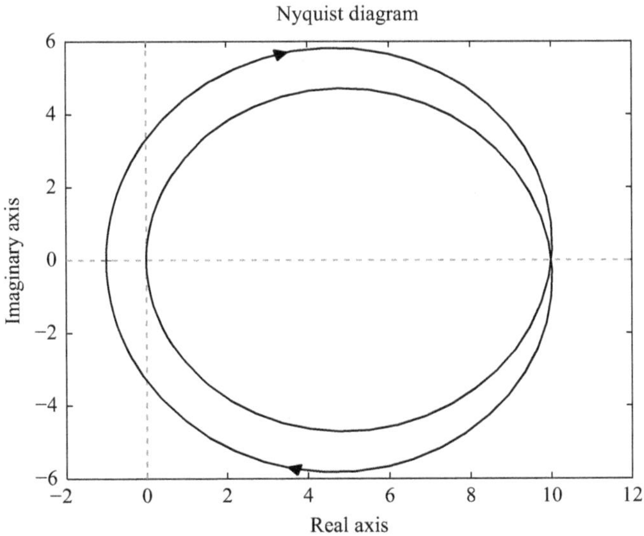

Figure 5.7 Nyquist plot of $\frac{100(s-1)}{s^2+10s+100}$

intersects the critical point. If no additional loop shaping is applied (only gain changes implemented), then the loop gain must be reduced to satisfy the Nyquist Criterion. This is an example of the limiting feature of RHP zeros in feedback systems, particularly the upper bound on proportional feedback they present (root locus analysis readily identifies this as well). The negative effect of an RHP zero is most pronounced in the neighborhood of its natural frequency. For $T(s)$, the phase change at $\omega << 1$ rad/s is very small (the phase contribution is approximately 180 degrees). Similarly, the phase change at $\omega >> 1$ rad/s is very small, and the contribution is ~90 degrees, which is the same as that of a LHP zero at frequencies much higher than the natural frequency. The RHP and LHP zero is indistinguishable at these frequencies from a phase perspective. In the neighborhood of the natural frequency, the RHP zero phase decreases by 90 degrees. It is this non-minimum phase delay that is a threat to stability and must be carefully taken into consideration when designing the controller. Bandpass-type loop transmissions where negative feedback is applied at frequencies much different than the RHP zero natural frequency are not adversely affected.

The Nyquist plot of $T(s) = \frac{100(s+1)}{s^2+10s-100}$ is shown in Figure 5.8. There is one unstable open-loop pole, and one anticlockwise encirclement of the critical point is required for closed-loop stability. An increase in loop gain accomplishes this. Now consider a similar system with T-plane plot shown in Figure 5.9. The poles are in the same location, the zero is at $s = 1$. The RHP zero reverses the direction of the Nyquist locus, and thus no combination of sign inversion or gain adjustment can provide the required anticlockwise encirclement of the critical point.

Figure 5.10 shows the T-plane plot of $T(s) = \frac{100}{s(s+10)}$. A condition of Cauchy's Theorem is that the *s-plane* contour not pass through any poles or zeros of the

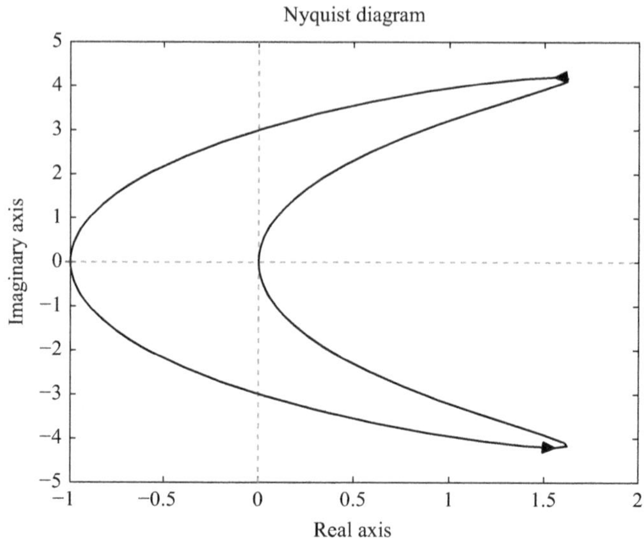

Figure 5.8 Nyquist plot of $\frac{100(s+1)}{s^2+10s-100}$

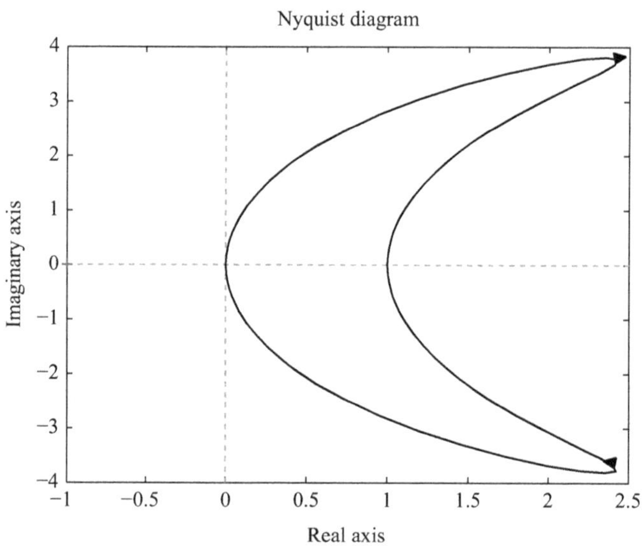

Figure 5.9 Nyquist plot of $\frac{100(s-1)}{s^2+10s-100}$

mapping function. The Nyquist contour does pass through the origin pole of $T(s)$. The modified Nyquist contour is implemented here, where poles and zeros on the $j\omega$ axis are circumvented by a semicircle of infinitesimal radius to their right. As such, these poles and zeros are not encircled by the s-plane contour and do not contribute to integers Z and P.

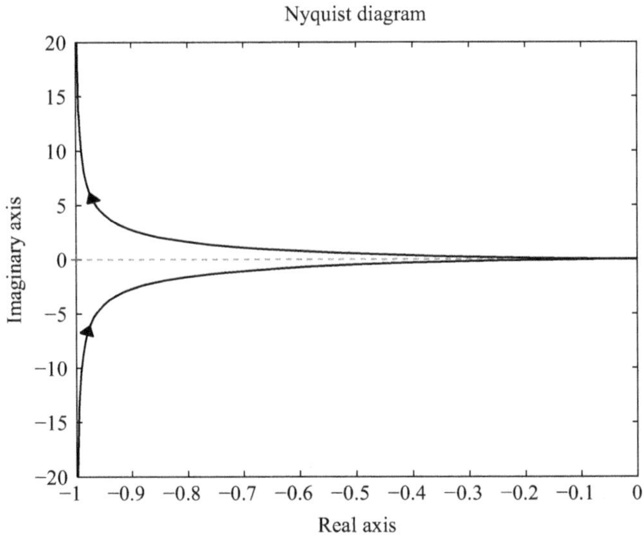

Figure 5.10 Nyquist plot of $\frac{100}{s(s+10)}$

The proximity of the modified Nyquist contour to the imaginary poles results in very large magnitude loci in the *T*-plane, making impossible the plotting of the entire *T*-plane locus. The plots are typically zoomed in about the origin, with breaks in the plot in the neighborhood of imaginary poles. This requires analysis of function *T*(*s*) to determine in which direction the unplotted curves of *T*(*s*) are rotating to determine the number of net encirclements of the critical point and the determination of stability of the closed-loop system using the Nyquist Criterion.

The plot starts slightly to the left of the negative imaginary axis (third quadrant) (first mapped point is $s = \epsilon$, ϵ an infinitesimal positive number) and approaches the origin along the negative real axis as $s \to j\infty$. The locus proceeds to the left of the positive imaginary axis as *s* takes on values on the negative real axis approaching the point $s = -j\epsilon$. Critical information related to closed-loop stability is missing in the *T*-plane plot. Are there encirclements of the critical point by the plot not shown in the figure? The critical missing part of the Nyquist plot is the mapping of the semicircular arc around the origin pole of *T*(*s*). The modified Nyquist contour starting at $s = -j\epsilon$ and ending at $s = j\epsilon$ has a phase increase of π radians. The mapping of this through *T*(*s*) results in a phase change of $-\pi$ in the *T*-plane (origin root of the *denominator* polynomial). This is a rotation through the first and fourth quadrants of the *T*-plane. As such, the Nyquist plot does not encircle the critical point, and the closed-loop system is stable (no poles of *T*(*s*) in RHP).

Consider the loop transmission function $T(s) = \frac{10}{s(s+1)^2}$, mapping the modified Nyquist contour to the complex function plotted in Figure 5.11. The Nyquist Criterion requires no net encirclements of the critical point (*T*(*s*) has an origin pole and

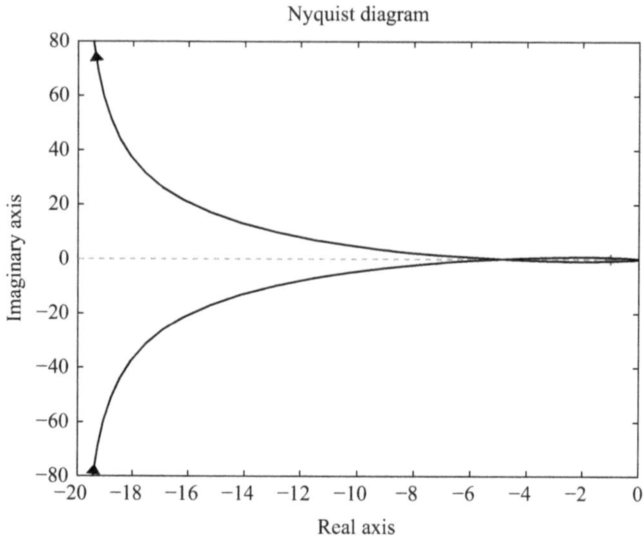

Figure 5.11 Nyquist plot of $\frac{10}{s(s+1)^2}$

two LHP poles, $P=0$). Again the plot starts along the negative imaginary axis, but as the relative degree is 3, the Nyquist plot approaches the origin along the positive imaginary axis as $s \rightarrow j\infty$. The remainder of the plot is symmetric to this as s approaches the semicircle about $s=0$ along the negative imaginary axis of the s-plane. There is one clockwise encirclement of the critical point seen in this plot. It is important to determine if there are any more encirclements of the unplotted part of the Nyquist plot to determine if the closed-loop system is stable (an unplotted counterclockwise encirclement, making the number of net encirclements zero) or unstable (no additional encirclements or additional clockwise encirclements).

Like the previous example, there is a single origin pole of $T(s)$. The phase in the T-plane is $-\pi$, and the unplotted part of the Nyquist locus wraps clockwise from the top to the bottom of the plot. Thus, a second clockwise encirclement is completed, and there are $N=2$ unstable poles in the closed-loop system.

Consider the Nyquist plot of $T(s)=\frac{1}{s^2(s+10)}$ shown in Figure 5.12. The function has one LHP pole and two origin poles. The phase of the system at low frequency is slightly less than 180 degrees, and the plot approaches the origin along the positive imaginary axis (relative degree is 3). The plot then proceeds along the negative real axis in the third quadrant as s takes on negative imaginary values. There are two origin poles, so the phase change in the T-plane from the mapping of the modified Nyquist semicircle about the origin is 2π. The T-plane plot wraps around the origin in the clockwise direction, and the critical point is encircled twice clockwise. There are two unstable closed-loop poles.

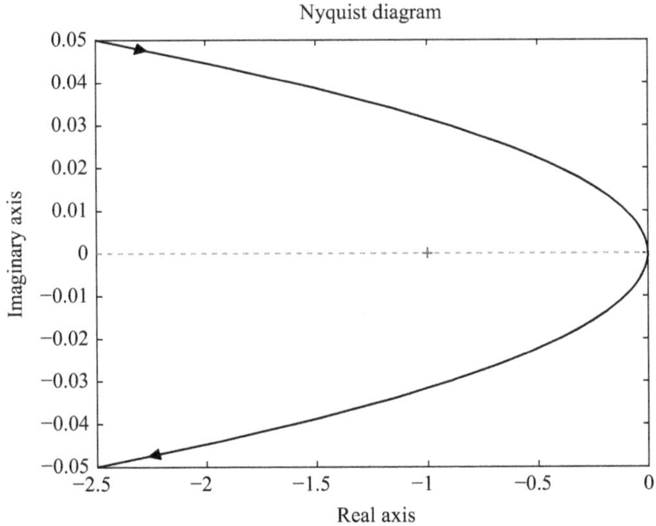

Figure 5.12 Nyquist plot of $\frac{1}{s^2(s+10)}$

5.4 Relative stability

The Nyquist Stability Criterion indicates whether a closed-loop system is stable or unstable. In Chapter 3, it is stated that positive feedback ($|F(s)| < 1$) amplifies disturbances. Consider two control systems with bandwidth ω_1. Both systems are stable in the open-loop condition, and the T-plane plots have zero net encirclements of the critical point. Both systems are stable by the Nyquist Stability Criterion. The loop transmissions at crossover are $T_1(j\omega_1) = -0.5 - j0.866$ and $T_2(j\omega_1) = -0.996 - j0.087$, respectively. The *feedback* $|1 + T(j\omega_1)|$ is substantially different, 1 and 0.0872, respectively, so for a sinusoidal disturbance at ω_1 additive at the output of the plant, the first system does not amplify this disturbance, whereas the second system amplifies it by a factor of more than 11. Although both controllers are stable by the Nyquist Stability Criterion, the first controller is *relatively* more stable than the second. It is evident that there is a relationship between positive feedback and relative stability.

Measures of relative stability relate to the closeness of the T-plane Nyquist plot to the critical point. The *phase margin* in degrees is a measure of closeness to the critical point of the T-plane plot as it crosses the unit circle, $|T| = 1$.

$$\phi_m = -(-180^\circ - arg(T(j\omega))_{|T(j\omega)=1|}) \tag{5.35}$$

The larger the phase margin, the greater the separation of the T-plane plot on the unit circle and the critical point, and the greater the relative stability. The phase margins of the two example systems are 60 and 5 degrees, respectively. If the T-plane plot crosses the unit circle in the third or fourth quadrant, the phase margin is positive. If it crosses in the first or second quadrant, the phase margin is negative. The magnitude of the phase margin indicates closeness to the critical point, while the sign gives an indication of whether or not the T-plane plot is wrapping around the critical point.

Another measure of relative stability is the *gain margin* in decibels, which measures the separation of the T-plane plot as it intersects the negative real axis and the critical point.

$$K_m = -20\log_{10}x \tag{5.36}$$

Quantity x is the inverse of the magnitude of T as it crosses the negative real axis. So, if the T-plane plot crosses the negative real axis inside the unit circle, the gain margin is positive. If it crosses outside the unit circle, the gain margin is negative. Like the sign of the phase margin, the sign of the gain margin indicates if the T-plane plot is wrapping around the critical point.

For control systems that are stable in the open loop, phase and gain margin give a measure of how much increase in the loop gain and/or decrease in the loop phase is acceptable before stability is threatened because they give an indication of how close the T-plane plot is to encircling the critical point. There are instances where the utility of these measures is reduced. The T-plane plot of a controller that is unstable in the open loop must encircle the critical point in the anticlockwise direction, so negative margins of stability are possible, but necessary for satisfaction of the Nyquist Stability Criterion. Stability margins for advanced control designs might be misleading. Figures 5.13 and 5.14 are the T-plane plots of a controller stable in the open loop. Clearly, the system is stable by the Nyquist Stability Criterion, but the phase margin is negative, thereby giving a false

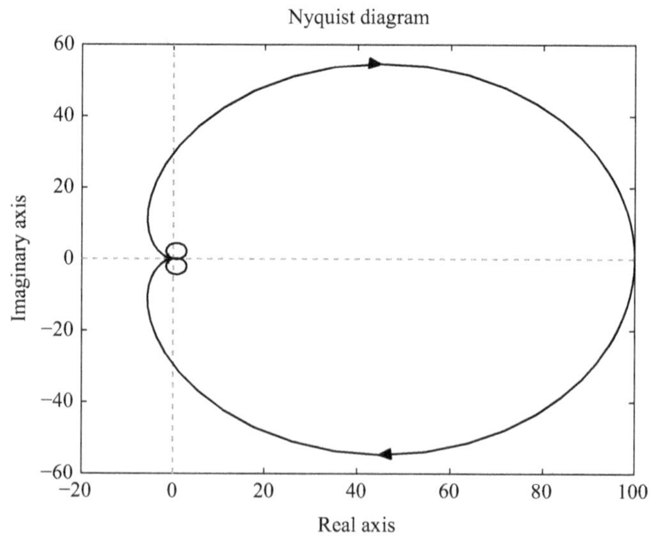

Figure 5.13 A stable controller with a negative phase margin

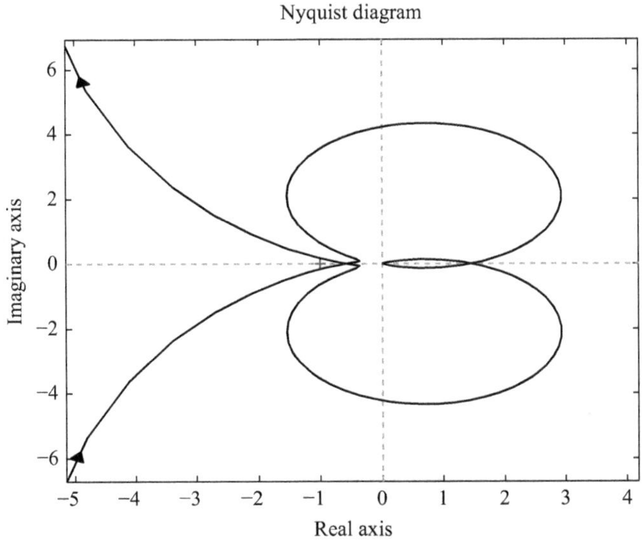

Figure 5.14 A stable controller with a negative phase margin (zoomed in around origin)

indication of poor relative stability. This type of plot is common in *phase-stabilized* controllers, which will be discussed later. Figures 5.15, 5.16 and 5.17 are the *T*-plane plots of a controller stable in the open loop. There are three crossings of the

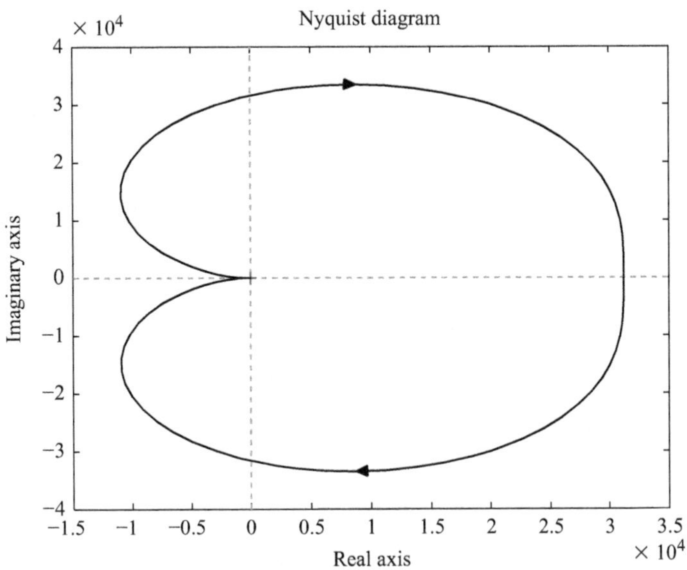

Figure 5.15 A stable controller with a negative gain margin

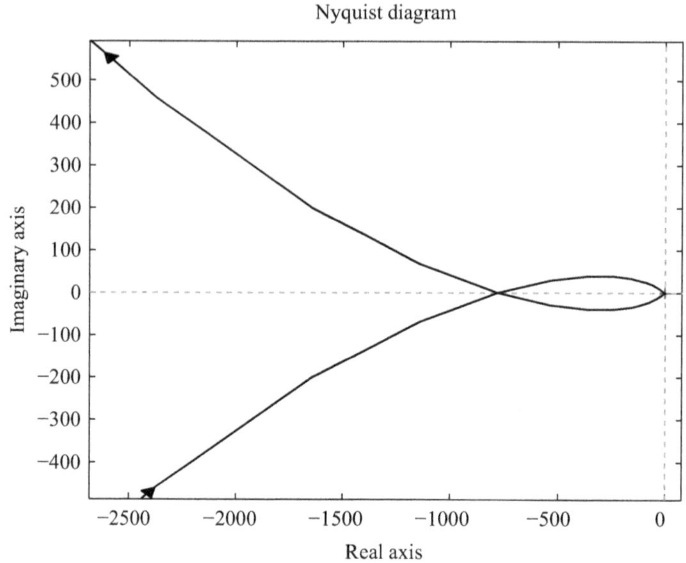

Figure 5.16 A stable controller with a negative gain margin (zoomed in around the origin)

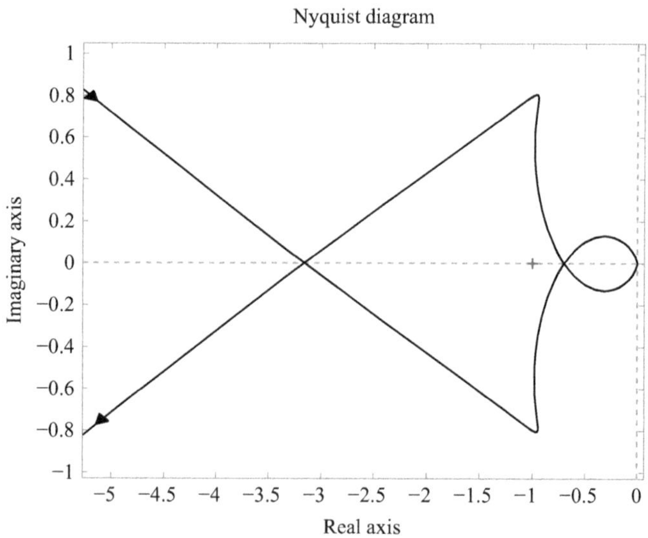

Figure 5.17 A stable controller with a negative gain margin (zoomed in closer to the origin)

negative real axis: two outside the unit circle (negative gain margins) and one inside. Is the system stable? There are zero net encirclements of the critical point, so the closed-loop system is stable by the Nyquist Stability Criterion. Yet how are the multiple negative gain margins to be interpreted? In this case, the negative gain margins give an indication of how much gain *reduction* is possible before the critical point is encircled, not an indication that the critical point is being encircled. This is a *Nyquist-stable* controller and will be discussed later.

It is evident that there is a relationship between the margins of stability and positive feedback. Small margins indicate a closeness to the critical point, or $|1 + T(j\omega)|$ is small (positive feedback). So, controllers with poor relative stability will amplify disturbances at frequencies close to the crossover.

5.5 Internal stability

A transfer matrix is exponentially stable if and only if it is proper (improper systems may not be BIBO stable) and has no poles in the closed right half plane (CRHP). Consider the feedback system shown in Figure 5.18. The mapping of the external signals to internal signals is

$$\begin{bmatrix} e_1(s) \\ e_2(s) \end{bmatrix} = \begin{bmatrix} H_{11}(s) & H_{12}(s) \\ H_{21}(s) & H_{22}(s) \end{bmatrix} \begin{bmatrix} u_1(s) \\ u_2(s) \end{bmatrix} = H(s) \begin{bmatrix} u_1(s) \\ u_2(s) \end{bmatrix} \tag{5.37}$$

The feedback system is said to be *internally stable* if and only if the transfer matrix $H(s)$ is exponentially stable. Satisfaction of this definition eliminates the possibility of RHP pole-zero cancelations between systems $G_1(s)$ and $G_2(s)$ that stability analysis using the loop transmission matrix $G_1(s)G_1(s)$ cannot account for. A system with RHP pole-zero cancelations might be stable in an input–output sense, but internal signals yield unbounded outputs.

If both $G_1(s)$ and $G_2(s)$ are unstable, then all four transfer matrices of $H(s)$ need to be checked for exponential stability. However, if one of the two, say $G_2(s)$ is stable, then only $H_{21}(s) = [I - G_1(s)G_2(s)]^{-1}G_1(s)$ need to be checked for exponential stability. This is a common condition, as while the plant in the forward path might cause $G_1(s)$ to be unstable, systems in the feedback path are typically related to the sensors and are stable. Furthermore, if $G_2(s)$ is exponentially stable, $H_{21}(s)$ is exponentially stable if and only if $det[I - G_1(s)G_2(s)]$ has no zeros in the CRHP and $[I - G_1(s)G_2(s)]^{-1}G_1(s)$ is analytic at every CRHP pole of $G_1(s)$. The first

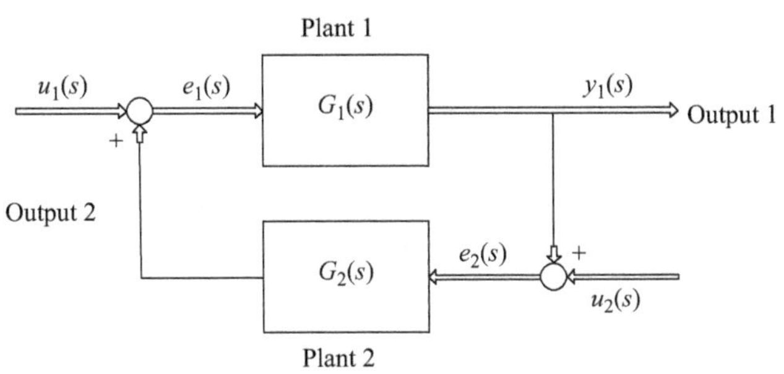

Figure 5.18 Feedback system with two external inputs

condition is no poles of the closed-loop system in the CRHP, and the second condition prevents CRHP pole-zero cancelation.

5.6 Generalized Nyquist Stability Criterion

Stability of the multivariable feedback system can be ascertained by analysis of the loop transmission matrix $kT(s)$, where $k \in \Re$, $T(s) \in \mathbf{F}^{n\times n}(s)$. The function $det[I + kT(s)]$ has p poles and z zeros in the CRHP. By the Principle of the Argument, the change in argument of this function as the Nyquist contour is mapped through it is equal to the difference in CRHP poles and zeros multiplied by 2π.

$$\Delta arg(det[I + kT(s)]) = -2\pi(z - p) \tag{5.38}$$

A condition of internal stability is $det([I + kT(s)])$ having no CRHP zeros ($z = 0$), so $\Delta arg(det[I + kT(s)]) = 2\pi p$. Unlike the case of SISO Nyquist analysis, where the effect of increasing or decreasing k on the Nyquist plot is clear, an analysis of the plot must be performed for every value of k. Consider the eigenvalue decomposition of matrix $T(s) = M(s)\Lambda(s)M^{-1}(s)$, where $M(s)$ and $\Lambda(s)$ are the eigenvector matrix and the diagonal matrix of eigenvalues as a function of the Laplace variable, respectively. A single gain is applied to all channels, so $kT(s) = kIM(s)\Lambda(s)M^{-1}(s) = M(s)[k\Lambda(s)]M^{-1}(s)$. $I + kT(s) = M(s)M^{-1}(s) + M(s)k\Lambda(s)M^{-1}(s) = M(s)[I + k\Lambda(s)]M^{-1}(s)$. The determinant is the product of the eigenvalues, so $det[I+kT(s)] = (1+k\lambda_1(s))(1+k\lambda_2(s))\ldots(1+k\lambda_n(s))$ and $\Delta arg(det[I + kT(s)]) = \sum_{i=1}^{n} \Delta arg[1 + k\lambda_i(s)]$. Closed-loop stability is now assessed by counting the number of encirclements of the origin by the plots of $1 + k\lambda_i(s)$, or equivalently the number of encirclements of the critical point $(0 + j1)$ by the plots of $k\lambda_i(s)$. These plots are referred to as the *characteristic loci*.

5.7 Gershgorin analysis

The Generalized Nyquist Stability Criterion requires the determination of the eigenvalues of the loop transmission matrix over a large range of frequencies, and the plotting of the characteristic loci to determine the stability of the multivariable feedback system. There is an elegant theorem developed by the Byelorussian mathematician Gershgorin that can be used as an alternative to this that does not require eigenvalue calculation.

Gershgorin's Theorem
The eigenvalues of an $n \times n$ complex matrix T lie in the union of n circles. The center of the ith circle $(1 \le i \le n)t_{ii}$ is the ith diagonal element of T. The radius of this circle is

$$\sum_j t_{ij}x_j = t_{ij}$$

The eigenvalues also lie in the union of circles, each with center t_{ii} and radius $\sum_{i \neq j} |t_{ij}|$.

Proof
Let λ be an eigenvector of matrix T and x be the corresponding eigenvector. Let x_i have the largest absolute value in x. $|x_i| > 0$, or $x = 0$.

$$Tx = \lambda x \tag{5.39}$$

$$\sum_j t_{ij} x_j = \lambda x_i \tag{5.40}$$

$$\sum_{j \neq i} t_{ij} x_j = \lambda x_i - t_{ii} x_i \tag{5.41}$$

Divide both sides by x_i.

$$|\lambda - t_{ii}| = |\frac{\sum_{j \neq i} t_{ij} x_j}{x_i}| \leq \sum_{j \neq i} |t_{ij}| \tag{5.42}$$

Gershgorin's Theorem is applied to the problem of determining multivariable feedback control stability.

Definition: Nyquist array
An array of individual Nyquist plots, the *ij*th being the Nyquist plot of the *ij*th transfer function of the transfer matrix.

Consider the Nyquist array of square transfer matrix $T(s)$. Superimpose on each point of the complex plot of $t_{ii}(j\omega)$ a circle of radius $\sum_j |t_{ij}|, j \neq i$ or $\sum_i |t_{ij}|, i \neq j$. The circles create bands, referred to as Gershgorin bands, the union of which contains the union of the characteristic loci. It can be shown that if the Gershgorin bands occupy distinct regions, as there are as many characteristic loci contained in the region as the number of Gershgorin bands occupying it (see Maciejowski). Thus, if all Gershgorin bands of the loop transmission matrix exclude the critical point, one can assess closed-loop stability by counting the encirclements of the critical point by the Gershgorin bands. This allows determination of control system stability without having to calculate the eigenvalues of the transfer matrix evaluated over the Nyquist contour.

It is noted that the Gershgorin Theorem applied to multivariable stability determination is a sufficient condition. If the Gershgorin bands overlap the critical point, nothing can be stated about the stability of the feedback system.

5.7.1 Case study: multiaxis control of a parallel robot

As opposed to serial robots, the base of the robot and the end-effector form a closed kinematic chain. These types of robots have increased payload and speed and are potentially more accurate than their serial counterparts, with the drawbacks of small workspace and more complicated kinematics and dynamics. A diagram of

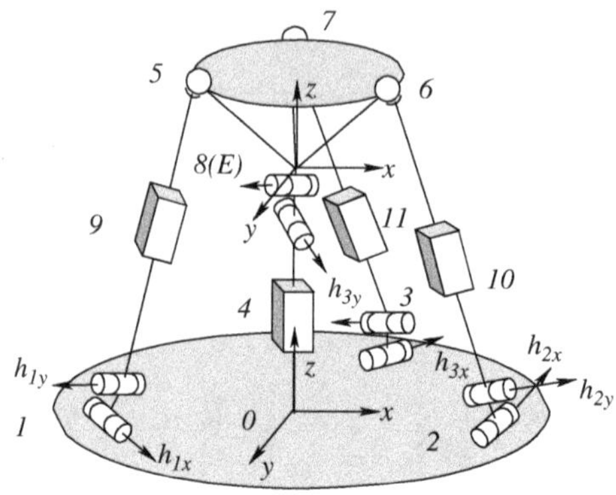

Figure 5.19 3-UP̲S – 1 – PU parallel robot

the 3 – *UP̲S*, 1 – *PU* is shown in Figure 5.19. Points 1, 2, 3 and 4 are fixed on the static *base*. Points 5, 6, 7 and 8(E) are fixed on the moving, rigid *end-effector*. The base and the end-effector are connected by four legs. The leg that connects points 1 and 5 has the following architecture: a passive universal joint at point 1 with revolute axes h_1x and h_1y, an active prismatic joint that changes the distance from points 1 and 5, and a passive spherical joint at point 5. The same architecture is applied to the legs connecting points 2 and 6, and 3 and 7. The vertical leg located at the center of the circle defined by points 1, 2 and 3 on the circumference has the following architecture: a rigid connection to the base, a passive prismatic joint (slider) connecting points 0 and 8(E), and a passive universal joint at point 8(E) with revolute axes x and y. A reference frame is fixed to the base with its origin at point 0 (the $\{0\}$ frame) and another is fixed to the end-effector with its origin at point 8(E) (the $\{E\}$ frame). The *pose* of the end-effector is defined at the position of 8(E) and the orientation of frame $\{E\}$ referenced to the position of 0 and the orientation of frame $\{0\}$, respectively.

The 3 – *UP̲S*, 1 – *PU* is a *limited-DOF* parallel mechanism, defined as a spatial mechanism with fewer than 6 active DOFs. A wrench applied on the end-effector is resisted by a combination of active joint forces/torques and the robot structure. The 3 – *UP̲S*, 1 – *PU* center leg allows only translation in the z-direction, and rotation in the x- and y-directions. The passive joints at the ends of the active legs allow the transmission of forces along the leg axes only. The remaining force/torque is supplied by the structure of the center leg (x- and y-forces and z-torques). It is clear that the center leg must be sufficiently sturdy to adequately provide this resistive force/torque. A unique design feature of this parallel mechanism is the use of high-force voice actuation for high-bandwidth control of heavy payloads with simultaneous passive isolation. The actuators are compliant high-force voice coils. The decade 1–10 Hz is targeted for negative feedback, while high frequency disturbances are attenuated passively due to soft suspension.

Plant identification

To identify the plant frequency responses, a random signal is digitized and the inverse kinematic solution is applied to generate either tip or tilt axis excitation.

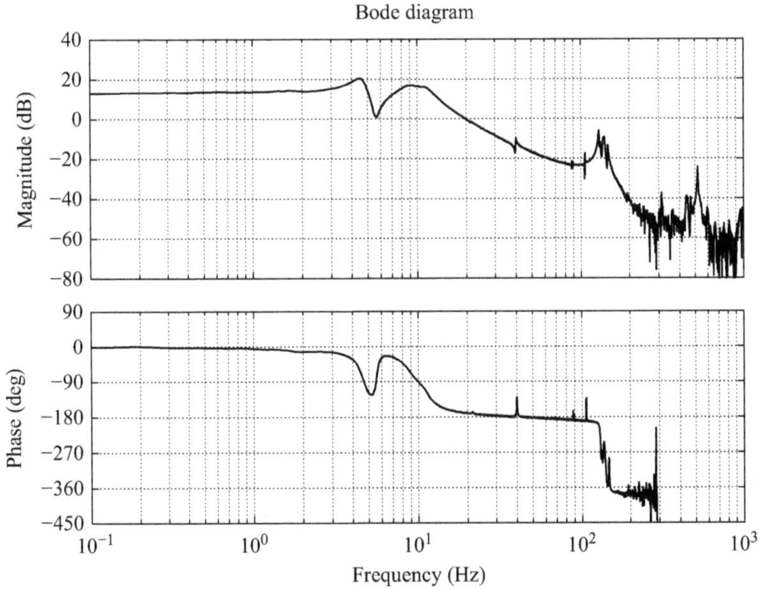

Figure 5.20 Plant frequency response: x-input/x-output

The three signals are converted to analog and sent to the three actuator current amplifiers. For each axis excitation, both tip and tilt measurements from the sensor are recorded. The diagonal frequency responses are shown in Figures 5.20 and 5.21. The off-diagonal frequency response moduli are shown in Figure 5.22. While there is coupling between axes caused by a deadzone in the $1 - PU$ universal joint, it is most significant in the neighborhood of 10 Hz, where the loop gain is largest and the Gershgorin circles are centered near their maximum distance from the critical point. The control problem is treated as two independent, single-input/single-output designs.

Compensator design

The control design goal is to apply as much feedback as feasible in the 1–10 Hz decade. Feedback bandwidth limitations include a second-order roll-off in plant response at frequencies higher than 10 Hz, plant resonances starting at 40 Hz, and sensor noise becoming restrictive in the 100–1000 Hz decade. Two fourth-order compensators are designed for the two-axis controller. The frequency response for the x-axis controller is seen in Figure 5.23. The y-axis compensator is similar. The conjugate zeros at 0.8 Hz boost the feedback in the critical frequency interval 1–10 Hz. A wide first-order lead provides phase margin for multiple return ratio 0 dB crossovers in the 10–100 Hz decade. The flexible body modes in the octave 100–200 Hz represent a severe performance limitation if gain is stabilized, and a robustness threat if the quality factors are time-varying. Thus, the modes are phase stabilized by a real pole in the compensator at 100 Hz. The return ratios for the controller are seen in

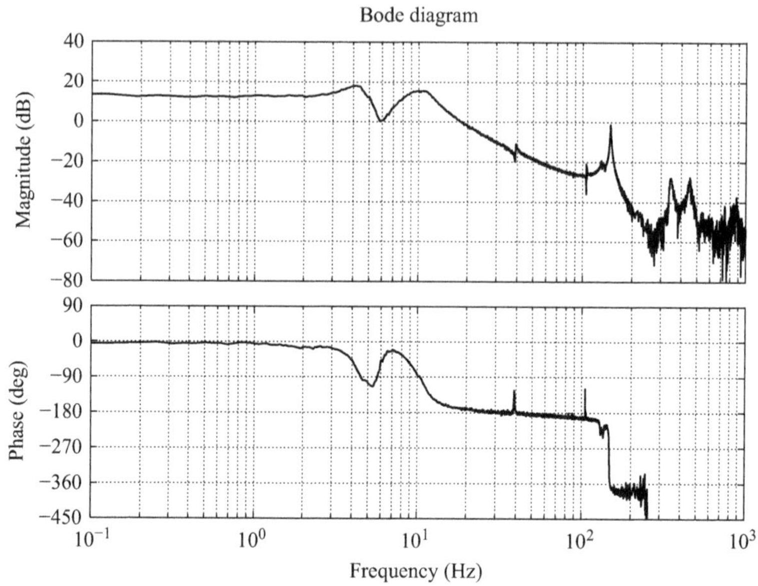

Figure 5.21 Plant frequency response: y-input/y-output

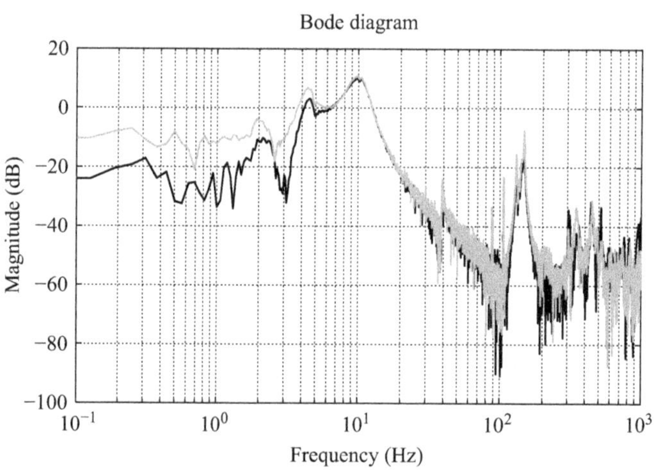

Figure 5.22 Plant frequency response: x-input/y-output, y-input/x-output

Figures 5.24 and 5.25. Maximum negative feedback in the 1–10 Hz decade is approximately 27 dB. Considering the feedback limitations, the controller is aggressive. If base disturbance signal power in the 10–100 Hz decade is large, the controllers will have to be modified slightly to increase stability margins at the cost of reduced low-frequency negative feedback.

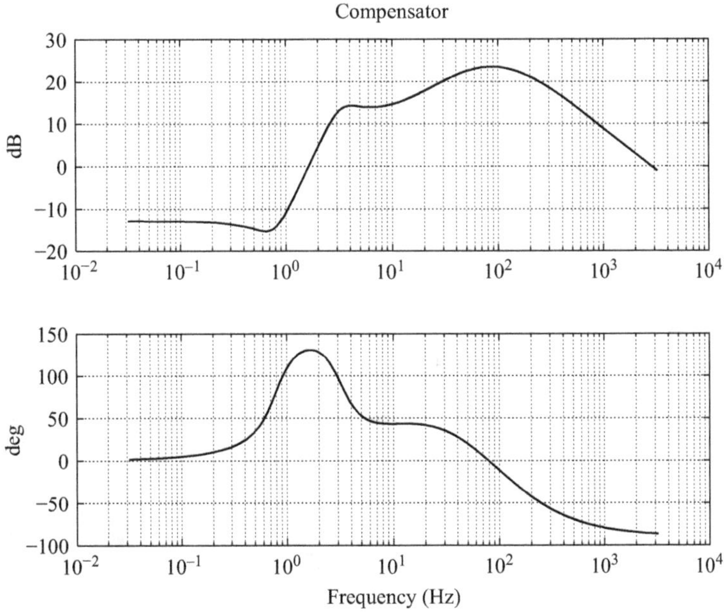

Figure 5.23 Vibration suppression compensator frequency response function

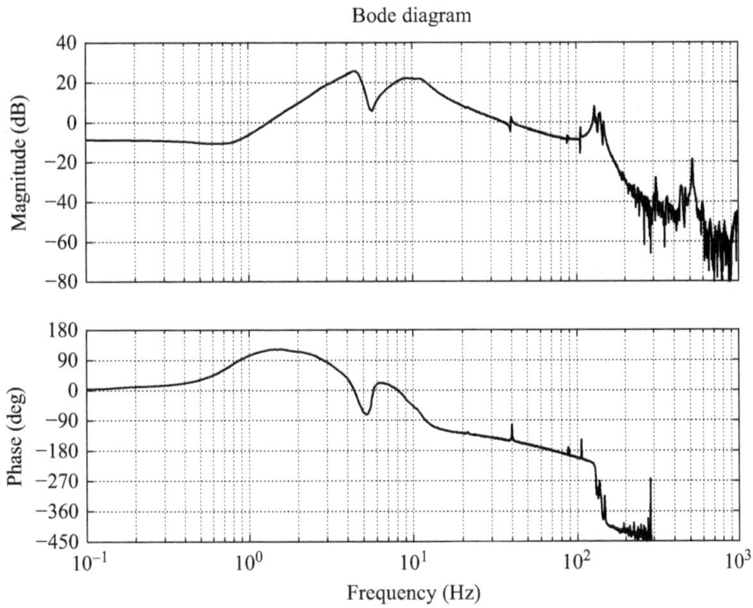

Figure 5.24 X-axis return ratio frequency response

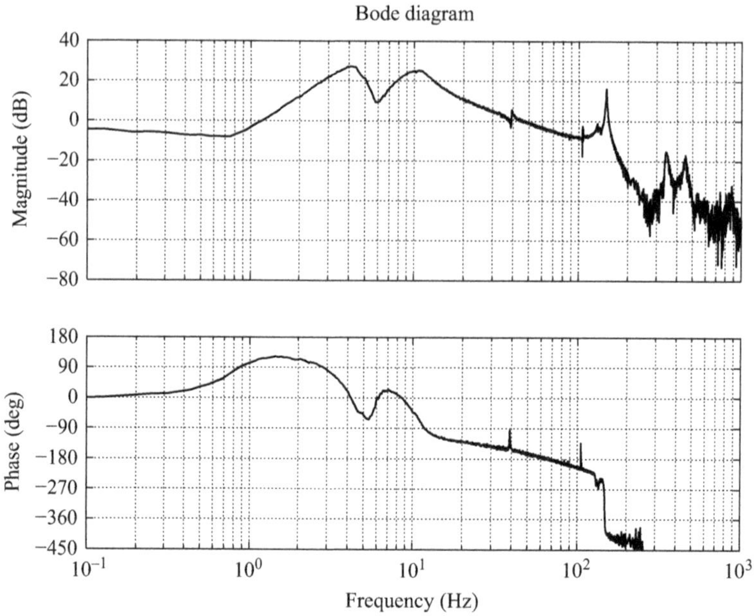

Figure 5.25 Y-axis return ratio frequency response

Figures 5.26 and 5.27 show the Gershgorin circles superimposed on the diagonal elements of the 2 × 2 Nyquist array for the vibration suppression controller return ratio. As expected, the largest diameter circles (cross-coupling at the rigid body mode frequencies) do not overlap the critical point. The circles clustered around the origin indicate cross-coupling due to the phase-stabilized flexible-body modes at 800–900 rad/s. The plots indicate a lack of diagonal dominance (the origin is overlapped), and the critical point is overlapped in the (2, 2) Nyquist element; however, as the test is only sufficient, this is not an indication of instability. Indeed, using the Generalized Nyquist Criterion for an open-loop stable multivariable system, the characteristic loci plot shown in Figure 5.28 indicates closed-loop stability.

5.8 Lyapunov method

It is desired to determine the stability of an equilibrium point ($x_e \ni f(x_e, t) \equiv 0$) of a nonautonomous system $\dot{x} = f(x(t), t)$, $x(t_0) = x_0 \in \Re^n$ that satisfies the conditions of existence and uniqueness. It is noted that the equilibrium point can be shifted to the origin, i.e. $x_e = 0$. For systems with multiple equilibrium points (e.g. the pendulum), the stability of each must be assessed by origin shifts.

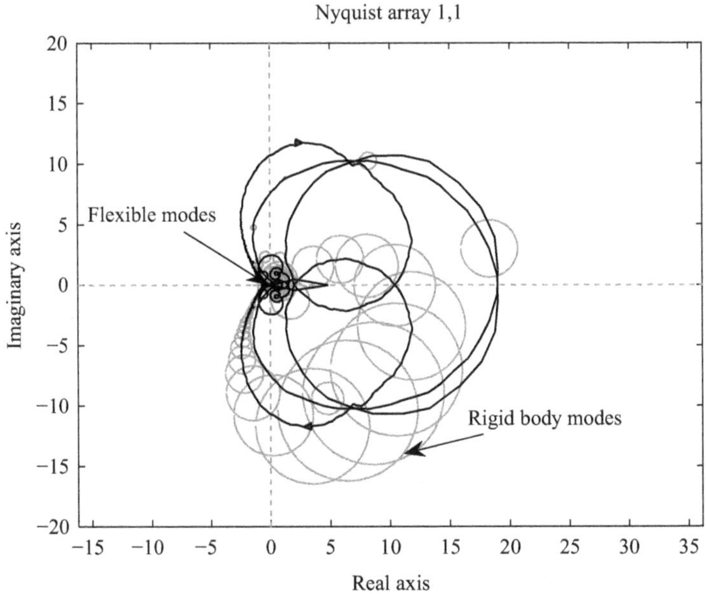

Figure 5.26 Gershgorin Circles on 1,1 Nyquist array plot

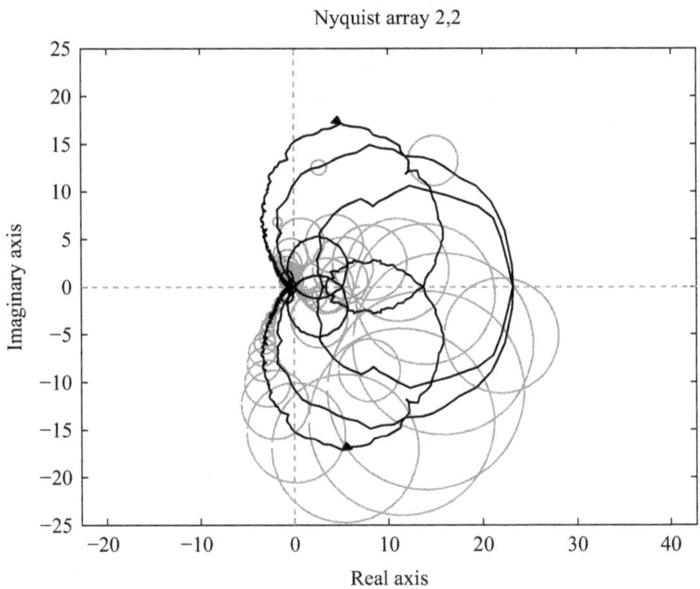

Figure 5.27 Gershgorin Circles on 2,2 Nyquist array plot

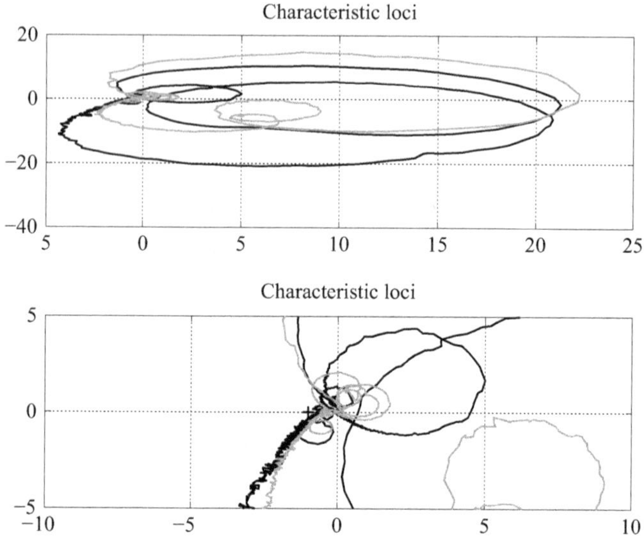

Figure 5.28 Characteristic loci for the vibration suppression controller

Definition: Stability in the sense of Lyapunov

The equilibrium point $x_e = 0$ of system $\dot{x} = f(x(t), t)$ is stable in the sense of Lyapunov at time $t = t_0$ if for any $\in > 0$, there exists *a* $\delta(t_0, \in) > 0$ such that $\|x(t_0)\| < \delta(t_0, \in) \Rightarrow \|x(t)\| < \in, \forall t \geq t_0$. If the condition is satisfied for δ independent of initial time t_0, the equilibrium point is said to be *uniformly* stable in the sense of Lyapunov.

The trajectory of a system that is stable in the sense of Lyapunov will stay within a ball around the origin if it starts within another ball around the origin. In other words, this type of system will not take off to infinity if it starts sufficiently close to the equilibrium point. This is analogous to the zero input response of a marginally stable linear system defined in Section 5.2, which is sometimes referred to as stable in the sense of Lyapunov. There is no guarantee of exponential, asymptotic, or any convergence to the origin. For most control applications, this characteristic is not acceptable as there is interest in driving the output error to zero.

Definition: Asymptotic stability

The equilibrium point $x_e = 0$ of system $\dot{x} = f(x(t), t)$ is asymptotically stable at time $t = t_0$ if x_e is stable in the Lyapunov sense and $\exists\ \delta(t_0) \ni \|x(t_0)\| < \delta(t_0) \Rightarrow \lim(x(t)) = 0$ as $t \to \infty$. If the equilibrium point is uniformly stable in the sense of Lyapunov, there exists δ satisfying the condition independent of t_0, and the convergence is uniform, then x_e is said to be uniformly asymptotically stable.

Clearly, asymptotic stability (particularly uniform asymptotic stability, where the stability of the equilibrium point is guaranteed) is a goal more appropriate for control applications than stability in the Lyapunov sense. However, there is no quantification of the convergence rate.

Definition: Exponential stability
The equilibrium point $x_e = 0$ of system $\dot{x} = f(x(t), t)$ is exponentially stable if $\exists$ constants a, $\alpha > 0$, $\in > 0 \ni ||x(t)|| \leq ae^{-\alpha(t-t_0)}||x(t_0)|| \;\forall\; ||x(t_0)|| \leq \in$ and $t \geq t_0$. The largest nonnegative constant α that satisfies this condition is the rate of convergence.

Autonomous systems that satisfy the stability condition are obviously uniformly stable. Systems that satisfy the equality conditions for all $x_0 \in \Re^n$ are *globally* stable.

5.9 Direct method

Consider a control problem where a function yielding a measure of error or energy in the system has been defined. Clearly, it is a good result if these quantities decrease with time. Lyapunov's direct method involves the determination of the rate of change of error or energy (power) to assess stability.

Definition: Locally positive definite function
Continuous function $V : \Re^n \times \Re^+ \to \Re$ is positive definite (locally) if in some neighborhood of the origin $x \in \{x \in \Re^n : ||x|| < \in\}$ $(\in > 0)$ and a continuous, strictly increasing function $a : \Re^+ \to \Re$, $V(0, t) = 0$ and $V(x(t), t) \geq a(||x(t)||)\ \forall t \geq 0$.

If this condition is satisfied with $a(||x(t)||) \to \infty$ as $||x(t)|| \to \infty$, the function is referred to as *globally postive definite*, or simply *positive definite.*

Definition: Locally negative definite (decrescent) function
Continuous function $V : \Re^n \times \Re^+ \to \Re$ is negative definite if in some neighborhood of the origin $x \in \{x \in \Re^n : ||x|| < e\}$ $(e > 0)$ and a continuous, strictly increasing function $b : \Re^+ \to \Re$, $V(x(t), t) \leq b(||x(t)||)\ \forall t \geq 0$.

Theorem: Lyapunov
Function $V(x(t), t)$ is a nonnegative function.

Local stability in the sense of Lyapunov
If $V(x(t), t)$ is locally positive definite and $\dot{V}(x(t), t) \leq 0$ locally in x, $\forall t$, then the origin is locally stable in the sense of Lyapunov.

Uniformly locally stable in the sense of Lyapunov
If $V(x(t), t)$ is locally positive definite and descresent, and $\dot{V}(x(t), t) \leq 0$ locally in x, $\forall t$ then the origin is uniformly locally stable in the sense of Lyapunov.

Uniformly locally asymptotically stable
If $V(x(t), t)$ is locally positive definite and descresent, and $-\dot{V}(x(t), t)$ is locally positive definite, then the origin is uniformly locally asymptotically stable.

Uniformly globally asymptotically stable
If $V(x(t), t)$ is positive definite and descresent, and $-\dot{V}(x(t), t)$ is positive definite, then the origin is uniformly globally asymptotically stable.

For an autonomous system, $\dot{x} = f(x)$, and continuously differentiable function, $V(x)$, mapping a domain D containing the origin to real numbers, if $V(0) = 0$, $V(x) > 0$ for all $x \in D \cap \bar{0}$ and $\dot{V}(x) \leq 0$ for all $x \in D$, then the origin $x = 0$ is stable in the sense of Lyapunov. If $\dot{V}(x) < 0$ for all $x \in D \cap \bar{0}$, the origin is asymptotically stable.

5.10 Case study: set point control of a parallel robot

Lyapunov stability theory is applied to the set point control problem for parallel robots. The goal of this controller is to transfer the robot from one pose (defined as the combination of position and orientation of the end-effector) to another. The trajectory of the robot as it transfers from the initial configuration to the desired set point is not directly controlled. Although the position of the end-effector is straightforward to describe in Euclidean 3-space ($p \in \Re^3$), the orientation is somewhat more complicated. Three-parameter (e.g. roll–pitch–yaw angles), four-parameter (quaternion), and matrix exponential representations of orientation can be used to describe orientation.

In this case study, a nonlinear kinematic set point control system is developed using the Lyapunov Theorem. This controller is applied to the *planar Stewart Platform* shown in Figure 5.29, and Seoul National University's *Eclipse* (Figure 5.30) parallel mechanisms.

5.11 Kinematic set point control

The velocity kinematics of a parallel robot are described by

$$\begin{bmatrix} \omega \\ \dot{p} \end{bmatrix} = \bar{J}_T \dot{\theta}_a \tag{5.43}$$

where ω and $\dot{p}$ are three-parameter variables in end-effector rotation and translation velocities, respectively; $\dot{\theta}_a$ is a vector of active joint velocities; and $\bar{J}_T$ is the composite manipulability Jacobian matrix found by solving for passive joint

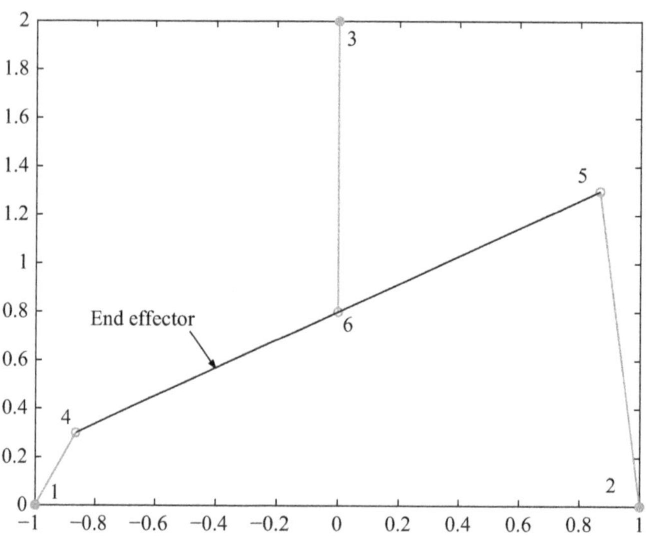

Figure 5.29 The Planar Stewart Platform

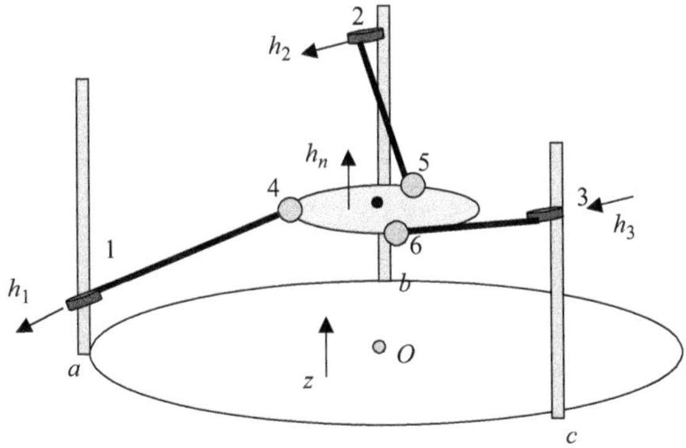

Figure 5.30 Seoul National University's Eclipse 5-face machining robot

velocities using the contact constraint equations. Using this mapping, a set-point kinematic control problem is proposed. Choose

$$\dot{\theta}_a \ni (R,p) \rightarrow (R_d,p_d) \tag{5.44}$$

The pairs (R, p) and (R, p) denote the initial and final configurations of the robot. The orientation is described by $R \in SO(3) = \{R \in \Re^{3\times 3} : RR^T = I,\ det(R) = 1\}$ and the position by $p \in \Re^3$. The end-effector error position is

$$e_p = \frac{1}{2}||p - p_d||^2 \tag{5.45}$$

Obtaining an error function for the rotational pose component is somewhat less straightforward. The following are several candidate functions:

$$e_R = \frac{1}{2}tr((\Sigma - I)^T)(\Sigma - I)) \tag{5.46}$$

$$e_R = \frac{1}{2}tr((\Sigma^{\frac{1}{2}} - I)^T)(\Sigma^{\frac{1}{2}} - I)) \tag{5.47}$$

$$e_R = \frac{1}{2}||P_\epsilon||^2 \tag{5.48}$$

$$e_R = \frac{1}{2}||q||^2 \tag{5.49}$$

where $\Sigma = R^T R_d \in SO(3)$, q is the vector quaternion, and P_e is any three-parameter representation of orientation (Euler angles, Gibb's vector, etc.). Using a three-parameter representation of orientation as in (5.48) is numerically expeditious, yet will always contain a representation singularity that can be confused with mechanism singularity.

A positive definite Lyapunov function is constructed via the sum of (5.45) and (5.49).

$$V = e_p + e_R \tag{5.50}$$

The time derivative of (5.50) is

$$\dot{V} = \Delta p^T \dot{p} + \frac{1}{2}\sin\frac{\phi}{2}\cos\frac{\phi}{2}k^T\omega = \Delta p^T \dot{p} + s^T\omega \tag{5.51}$$

where ϕ is the relative angle about the unit length axis of rotation k, and the 3×1 vector of reals, ω, is the task space angular rate. Applying the assumption that the mechanism is not at an unstable singularity (but could be in the close neighborhood), (5.51) is rearranged and combined with (5.43).

$$\dot{V} = \begin{pmatrix} s \\ \Delta p \end{pmatrix}^T \begin{pmatrix} \omega \\ \dot{p} \end{pmatrix} = \begin{pmatrix} s \\ \Delta p \end{pmatrix}^T \bar{J}_T \dot{\theta}_a \tag{5.52}$$

5.11.1 Kinematic control law

The manipulable variable, $\dot{\theta}_a$, is now driven by an appropriate control function. Candidate control laws are

$$\dot{\theta}_a = -\bar{J}_T{}^{\dagger} K_p \begin{pmatrix} s \\ \Delta p \end{pmatrix} \tag{5.53}$$

and

$$\dot{\theta}_a = -\bar{J}_T{}^{T} K_p \begin{pmatrix} s \\ \Delta p \end{pmatrix} \tag{5.54}$$

where K_p is a diagonal matrix of proportional gains. Applying (5.53) or (5.54) to (5.52) gives a negative definite quadratic function, thus both controllers are asymptotically stable with the stipulation that the mechanism is fully manipulable over the entire trajectory, i.e. $\bar{J}_T$ is well conditioned.

5.11.2 Examples of kinematic set point control

Set point control is applied to the Planar Stewart Platform and the Eclipse Manipulator. The trajectories do not include unstable singular configurations, which will be addressed in a later chapter.

Planar Stewart Platform

The PSP is tasked to move from $p_x = -0.5$, $p_y = 1$, $\theta = -\frac{\pi}{4}$, to $p_{x_d} = 0.5$, $p_{y_d} = 1.5, \theta_d = \frac{\pi}{4}$. The pseudoinverse control law (5.53) is implemented with $K_p = 50I$. The resulting trajectory and normed pose error are shown in Figures 5.31 and 5.32.

Eclipse Manipulator

The Eclipse is tasked to move from an initial pose (end-effector frame position and ZYX Euler angles relative to the base frame) $p = [0.1 \quad 0.05 \quad 0.01]^T, \theta = [0 \quad \frac{\pi}{12} \quad 0]$,

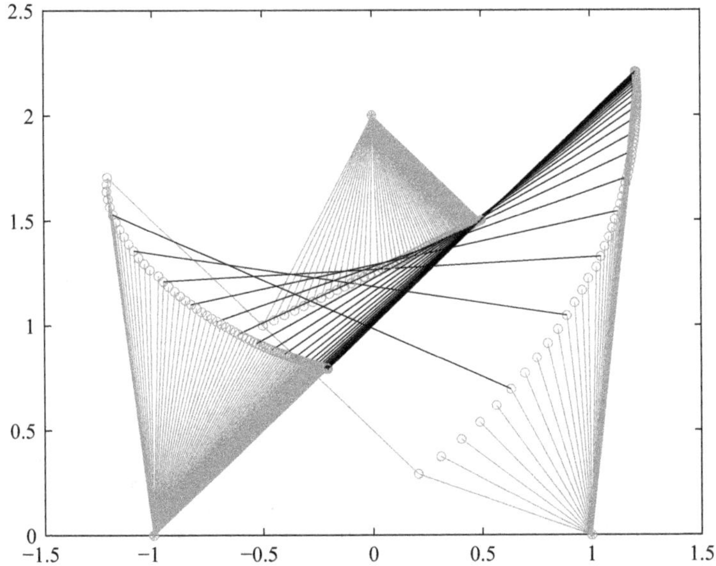

Figure 5.31 PSP trajectory: pseudoinverse set point control

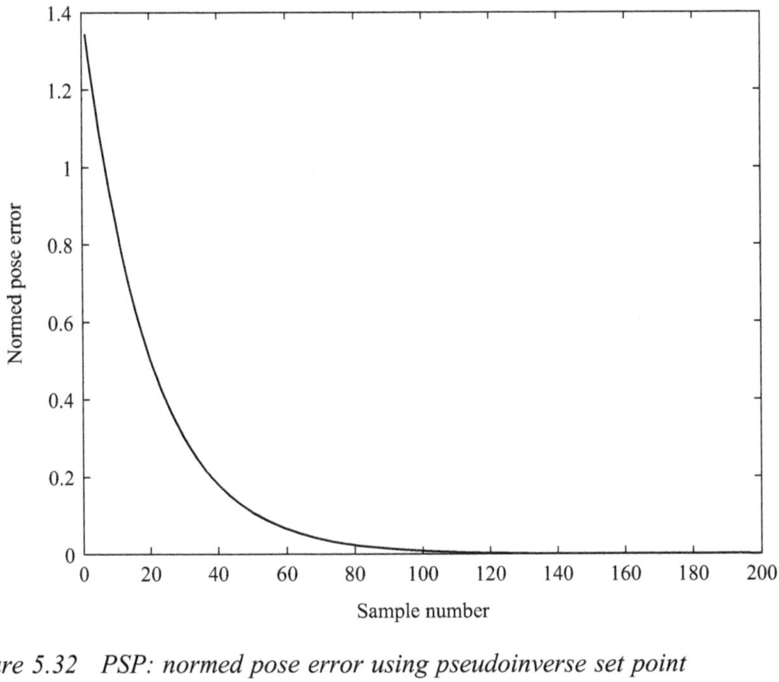

Figure 5.32 PSP: normed pose error using pseudoinverse set point control

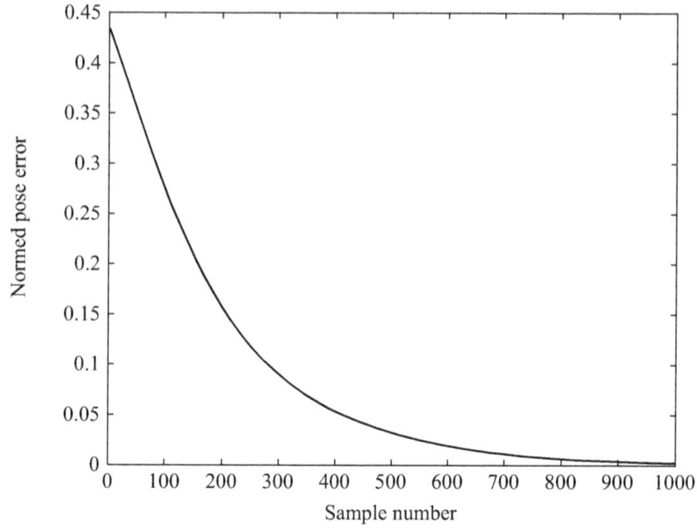

Figure 5.33 Eclipse: normed pose error with pseudoinverse set point controller

to $p_d = [-0.03 \quad 0.02 \quad 0.1]^T$, $\theta_d = [\frac{\pi}{6} \quad \frac{\pi}{6} \quad -\frac{\pi}{3}]^T$. Equation (5.53) is used with $K_p = 25I$. The normed pose error $\frac{1}{2}||\Delta p||^2 + \frac{1}{2}||q||^2$ is shown in Figure 5.33.

5.11.3 Effect of mechanism singularities on the kinematic set point controller

The pseudoinverse controller of (5.53) is ill-posed at unmanipulable singularity thus we consider the transpose controller of (5.54) in this section.

Unmanipulable singularity

At unmanipulable singularity, (5.52) is zero if the pose error is in the null space of the composite Jacobian. In this condition, the mechanism stops away from the set point.

Unstable singularity

At unstable singularity, (5.52) will have a nonzero drift term.

$$\begin{pmatrix} s \\ \Delta p \end{pmatrix}^T (\bar{J}_T \dot{\theta}_a + \tilde{\xi}) \tag{5.55}$$

The mechanism is uncontrollable at this pose. In the close neighborhood of unstable configurations, the maximum singular value of $\bar{J}_T{}^T$ becomes very large, and the magnitude of (5.54) becomes prohibitive.

5.12 Absolute stability

While linear feedback system stability theory is elegant and easy to implement, it does not address many aspects of control applications, perhaps most critically the effect of nonlinearities on the stability of the control system. Every control application involves actuators, and every actuator has limits (saturation). An obvious question is: If a linear control system is determined to be stable using the Nyquist Stability Criterion, what is the effect of particular nonlinearities in the loop on the system's stability?

Definition: Sector condition
Nonlinearity $\phi(t, y) : [0, \infty) \times \Re \rightarrow \Re$ satisfies the sector condition if $\exists a_1$, a_2 such that $a_1 y^2 \leq y\phi(t, y) \leq a_2 y^2$, $\forall t \geq 0$, $\forall y \in \Re$, where $a_2 > a_1$.

It might be the case that this sector condition is only satisfied locally in the interval $y \in (b_1, b_2)$, $b_1 < 0 < b_2$, the sector condition is said to be satisfied locally (as opposed to globally).

Definition: Sector condition (2)
Memoryless nonlinearity $\phi(t, y) : [0, \infty) \times \Re^p \rightarrow \Re^p$ satisfies the sector condition (said to belong to sector $[K_{min}, K_{max}]$) if

$$[\phi(t,y) - K_{min}y]^T[\phi(t,y) - K_{max}y] \leq 0, \forall t \geq 0, \forall y \in \Gamma \subset \Re^p \tag{5.56}$$

for real matrices K_{min}, K_{max} of sector lower and upper slopes (so $K = K_{max} - K_{min} > 0$). The set Γ has a connected interior that contains $y = 0$. If $\Gamma = \Re^p$, the nonlinearity belongs to this sector globally.

Consider a regulator (reference input is zero) consisting of a feedback connection of an LTI strictly proper system $G(s) = C(sI - A)^{-1}B$ in the forward path, and a nonlinear element $\phi(t, y)$ in the feedback path depicted in Figure 5.34. The feedback system is represented by

$$\dot{x}(t) = Ax(t) - B\phi(t,y) \tag{5.57}$$

$$y(t) = Cx(t) \tag{5.58}$$

It is assumed that the matrix pairs (A, B) and (A, C) are controllable and observable, respectively, and ϕ satisfies the sector condition.

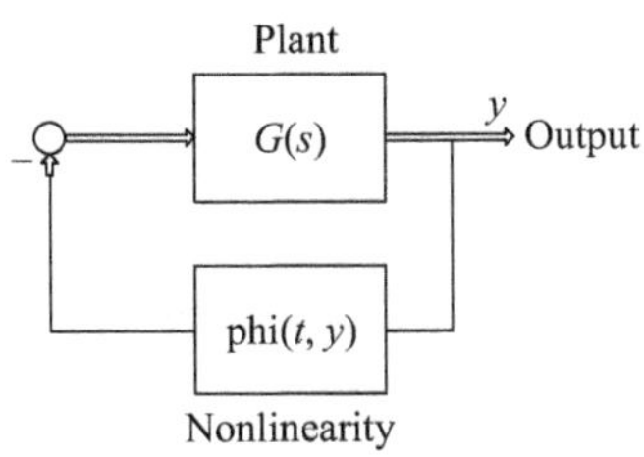

Figure 5.34 Linear system in feedback connection with a nonlinear system

Definition: Absolute stability
If the origin of system (5.57) is uniformly asymptotically stable for all nonlinearities in the sector that ϕ belongs to, the system is absolutely stable. If ϕ satisfies the sector condition globally, the feedback system is globally absolutely stable, otherwise it is absolutely stable on a finite domain.

Lemma: Positive Real Lemma
Consider square transfer matrix $Z(s) = C(sI - A)^{-1}B + D$, with all eigenvalues of A having negative real parts, and matrix pairs (A, B) and (A, C) controllable and observable, respectively. $Z(s)$ is strictly positive real if and only if there exists a positive definite symmetric matrix P, matrices W and L and positive constant ε such that

$$PA + A^T P = -L^T L - \varepsilon P \tag{5.59}$$

$$PB = C^T - L^T W \tag{5.60}$$

$$W^T W = D + D^T \tag{5.61}$$

See Khalil for the proof of this lemma, also referred to as the Kalman-Yakubovich-Popov Lemma.

5.12.1 Circle Criterion

Given system (5.57) where $\phi(t, y)$ satisfies the sector condition with $K_{min} = 0$, and all eigenvalues of A having negative real parts. In this case, the sector condition (5.56) is $\phi(t, y)^T\,[\phi(t, y) - KCx] \leq 0$, $\forall t \geq 0$, $\forall y \in \Gamma \subset \Re^p$, where $K > 0$ and symmetric.

A quadratic symmetric Lyapunov function is selected.

$$V(x) = x^T P x \tag{5.62}$$

where $P > 0$ is a symmetric matrix to be chosen later. Clearly, $V(x)$ is positive definite. The derivative of $V(x)$ is

$$\dot{V}(t,x) = \dot{x}^T P x + x^T \dot{P} x + x^T P \dot{x} \tag{5.63}$$

$$= (x^T A^T - \phi^T B^T) P x + x^T P(Ax - B\phi) \tag{5.64}$$

$$= x^T (PA + A^T P)x - 2x^T PB\phi \tag{5.65}$$

$$\leq x^T (PA + A^T P)x - 2x^T PB\phi - 2\phi^T(\phi - KCx) \tag{5.66}$$

$$= x^T (PA + A^T P)x + 2x^T (C^T K - PB)\phi - 2\phi^T \phi \tag{5.67}$$

The inequality is established by subtracting $2\phi^T\,(\phi - KCx) \leq 0$ from $\dot{V}$. The positive real lemma is now implemented to define a condition of absolute stability.

Given a symmetric, positive definite matrix P, matrix L and positive constant $\in$ such that

$$PA + A^T P = -L^T L - \in P \tag{5.68}$$

$$PB = C^T K - \sqrt{2}L^T \tag{5.69}$$

Substituting these into the inequality (5.66) yields the following:

$$\dot{V}(t,x) = -\in x^T Px - x^T L^T Lx + 2\sqrt{2}x^T L^T \phi - 2\phi^T \phi \tag{5.70}$$

$$= -\in x^T Px - (Lx - \sqrt{2}\phi)^T (Lx - \sqrt{2}\phi) \tag{5.71}$$

$$\leq -\in x^T Px \tag{5.72}$$

So, the derivative of the Lyapunov function is negative definite if matrices exist that satisfy $PA + A^T P = -L^T L - \in P$ and $PB = C^T K - \sqrt{2}L^T$. By the positive real lemma, this is the case if and only if $Z(s) = I + KC(sI - A)^{-1}B$ is strictly positive real.

Lemma

System (5.57), where all eigenvalues of A have negative real parts, and (A, B) and (A, C) are controllable and observable, respectively, with nonlinearity ϕ satisfying $\phi(t, y)^T\ [\phi(t, y) - KCx] \leq 0,\ \forall t \geq 0,\ \forall y \in \Gamma \subset \Re^p$ for symmetric, positive definite matrix K, is absolutely stable on a finite domain if $Z(s) = I + KC(sI - A)^{-1}B$ is strictly positive real. If $\Gamma = \Re^p$, the system is globally absolutely stable (or absolutely stable).

The requirement of open-loop stability can be removed from the lemma using *pole shifting* (see Khalil).

5.12.2 SISO case

If the linear system $G(s)$ is SISO ($p = 1$), the conditions of the Circle Criterion are the following:

1. If $0 < \alpha < \beta$, the Nyquist plot of $G(s)$ does not enter the disk whose diameter end points are $-\frac{1}{\alpha} + j0$ and $-\frac{1}{\beta} + j0$ and encircles this disk anticlockwise m times, where m is the number of open RHP poles of $G(s)$.
2. If $0 < \beta$, $G(s)$ is stable and the Nyquist plot of $G(s)$ lies to the right of the vertical line in the complex plane $-\frac{1}{\beta} + j\omega$, where $\omega \in (-\infty, \infty)$.
3. If $\alpha < 0 < \beta$, $G(s)$ is stable and the Nyquist plot of $G(s)$ is encircled by the disk whose diameter end points are $-\frac{1}{\alpha} + j0$ and $-\frac{1}{\beta} + j0$.

Example: Circle Criterion for a SISO system

The block diagram of a feedback system is shown in Figure 5.35. $C(s)P(s) = \frac{1000}{(s+1)(s+50)}$. The saturation is a model of the output limits of the actuator. The Nyquist plot of $C(s)P(s)$ is shown in Figure 5.36. As the Nyquist locus is to the right of the vertical line $s = -1$, the Circle Criterion is satisfied for the sector $[0 \quad 1]$ (saturation nonlinearity belongs to this sector).

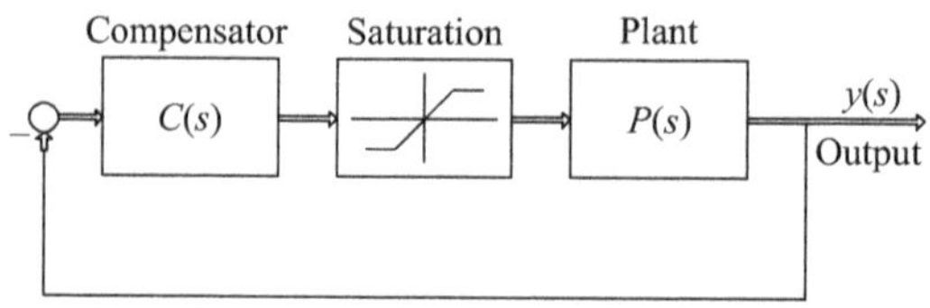

Figure 5.35 SISO feedback system with an actuator saturation

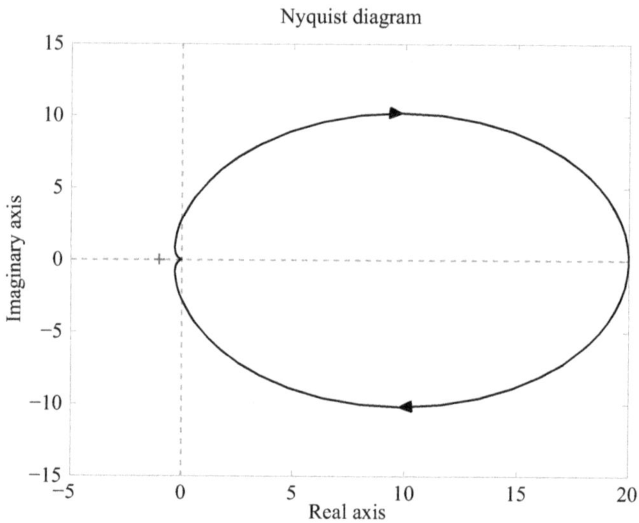

Figure 5.36 Nyquist plot of C(s)P(s)

5.12.3 Popov Criterion

Consider the linear system $\dot{x} = Ax - B\phi(y), y = Cx$ with all eigenvalues of A having negative real parts, and $\phi(y)$ a time invariant nonlinearity satisfying the sector condition $\phi(y)^T\,[\phi(y) - Ky] \leq 0,\ \forall y \in \Gamma \subset \Re^p$ where symmetric matrix $K > 0$. Furthermore, $\int_0^y \phi^T(\sigma)Kd\sigma \geq 0, \forall y \in \Gamma \subset \Re^p$. The Lure-type Lyapunov function is applied.

$$V(x) = x^T Px + 2\eta \int_0^y \phi^T(\sigma)Kd\sigma \tag{5.73}$$

where P is a symmetric, positive definite matrix, and η is a nonnegative constant to be chosen. A similar approach that is used in the prequel is applied to find the sufficient condition of absolute stability, the Popov Criterion.

Theorem

Consider the system $G(s) = C(sI - A)^{-1}B$ where all eigenvalues of A have negative real parts, (A, B) and (A, C) are controllable and observable, respectively, input is

$u = -\phi y$ where ϕ is a time invariant nonlinearity satisfying the sector condition globally with matrix $K > 0$, and $\int_0^y \phi^T(\sigma)Kd\sigma \geq 0, \forall y \in \Re^p$. Then the system is absolutely stable if there exists $\eta \geq 0$ whose negative inverse is not an eigenvalue of A such that $I + (1 + \eta s)KG(s)$ is strictly positive real. It is absolutely stable in a finite domain if the sector condition and $\int_0^y \phi^T(\sigma)Kd\sigma \geq 0$ are satisfied on a subset of $\Re^p$.

For the scalar case, $Z(s)$ is strictly positive real if and only if $Re[1 + (1 + j\eta\omega) kG(j\omega)] > 0, \forall\omega \in \Re$, equivalent to

$$\frac{1}{k} + Re[G(j\omega)] - \eta\omega Im[G(j\omega)] > 0, \quad \forall\omega \in \Re \tag{5.74}$$

If $Re[G(j\omega)]$ is plotted against $\omega Im[G(j\omega)]$, then this condition is satisfied if the plot lies to the right of the line with x-axis intercept $-\frac{1}{k}$ with slope $\frac{1}{\eta}$. This is the Popov plot. For $\eta = 0$, this condition becomes a Circle Criterion condition.

In introducing additional parameter, the Popov Criterion is less conservative than the Circle Criterion and thus might allow satisfaction of the absolute stability condition with wider sectors. This is traded with the ease of absolute stability analysis using the Circle Criterion, which in the SISO case requires nothing more than the readily available Nyquist plot of the equivalent linear system.

5.13 Exercises

1. The impulse response is $g(t) = 10e^{(-2.5t+5)}$. Is the system BIBO stable? Why?
2. The impulse response is $g(t) = -3e^{(-0.5t)} + 3.5te^{(-0.5t)}$. Is the system BIBO stable? Why?
3. The impulse response is $g(t) = 10e^{(-0.5t)} + 20\sin(20t)$. Is the system BIBO stable? Why?
4. The impulse response is $g(t) = 2.5e^{(0.5t)} + 3.5te^{(-0.5t)}$. Is the system BIBO stable? Why?
5. The impulse response is $g(t) = 10e^{(-0.5t)} + 3.5te^{(-0.5t)}$. Is the system BIBO stable? Why?
6. Is the system with transfer function $G(s) = 10\frac{s+1}{s^2+23s+40}$ stable in the strict sense, marginally stable, or unstable? Why?
7. Is the system with transfer function $G(s) = 20\frac{s-1}{s^2+23s+40}$ stable in the strict sense, marginally stable, or unstable? Why?
8. Is the system with transfer function $G(s) = 10\frac{s-1}{s^2+10}$ stable in the strict sense, marginally stable, or unstable? Why?
9. Is the system with transfer function $G(s) = 10\frac{s+1}{(s+10)(s^2+23s-10)}$ stable in the strict sense, marginally stable, or unstable? Why?
10. Report a bounded input to $G(s) = 10\frac{s-1}{s^3+16s}$ that yields a bounded output. Suggest two different bounded inputs that result in unbounded outputs. The difference in these signals should not simply be amplitude.

11. Determine if the linear system with state space realization (A, B, C, D) is stable.

$$A = \begin{bmatrix} 2.4 & 1-2 & \\ 1.1 & -8 & -11 \\ 2.3 & -1 & -10 \end{bmatrix}, B = \begin{bmatrix} 5 \\ 1 \end{bmatrix}, C = [3 \quad -2], D = 0$$

12. For the system described in the previous problem: Is (A, B) controllable? Is (A, C) observable? Is the state space realization minimal?

13. The state matrix is $A = \begin{bmatrix} 7.4 & 0 & -8-2 \\ 0 & -8 & -11 \\ 2.7 & -1 & -20 \end{bmatrix}$. Is the system zero input stable?

14. Two LTI systems are in cascade. $G_1(s) = 10\frac{s^2+5s-50}{s^2+25s+100}$ and $G_2(s) = 10\frac{s+85}{s^3+5s^2-29s-105}$. Find a state space realization (A, B, C, D) of this system. Is the realization minimal? Report any hidden modes. Is the system BIBO stable? Is it zero input stable? Is it internally stable?

15. The loop transmission function is $T(s) = K\frac{s+15}{s^2+25s+50}$. Use the Nyquist Stability Criterion to determine the set $K \in \Re$ for which the closed-loop system is stable.

16. The loop transmission function is $T(s) = K\frac{s-15}{s^2+25s+50}$. Use the Nyquist Stability Criterion to determine the set $K \in \Re$ for which the closed-loop system is stable.

17. The loop transmission function is $T(s) = K\frac{s+15}{(s+4)(s^2+25s+100)}$. Use the Nyquist Stability Criterion to determine the set $K \in \Re$ for which the closed-loop system is stable.

18. The loop transmission function is $T(s) = K\frac{-s-1}{s^2+25s-50}$. Use the Nyquist Stability Criterion to determine the set $K \in \Re$ for which the closed-loop system is stable.

19. The loop transmission function is $T(s) = K\frac{s+1}{s^2-25s+50}$. Use the Nyquist Stability Criterion to determine the set $K \in \Re$ for which the closed-loop system is stable.

20. The loop transmission function is $T(s) = K\frac{s-1}{s^2-25s-50}$. Use the Nyquist Stability Criterion to determine the set $K \in \Re$ for which the closed-loop system is stable.

21. The loop transmission function is $T(s) = K\frac{10}{s(s^2+25s+50)}$. Use the Nyquist Stability Criterion to determine the set $K \in \Re$ for which the closed-loop system is stable.

22. The loop transmission function is $T(s) = K\frac{s}{s^2+25s+50}$. Use the Nyquist Stability Criterion to determine the set $K \in \Re$ for which the closed-loop system is stable.

23. The loop transmission function is $T(s) = K\frac{s+2}{s^2+100}$. Use the Nyquist Stability Criterion to determine the set $K \in \Re$ for which the closed-loop system is stable.

24. Find the gain and phase stability margins for a feedback system with loop transmission function $T(s) = \frac{K}{s^2+20s+100}$ for $K = 10, 200, 2000$.

25. Find the gain and phase stability margins for a feedback system with loop transmission function $T(s) = 200\frac{(s^2+200s+10,000)}{(s+20)(s^2+10s+100)}$. Is the closed-loop system stable?
26. The loop transmission matrix is $T(s) = \begin{bmatrix} \frac{100}{s+20} & \frac{0.1}{s^2+2s+1} \\ \frac{1}{s+5} & \frac{500}{s^2+20s+100} \end{bmatrix}$. Determine the stability of the closed-loop system (if possible) using Gershgorin's Theorem. Determine the stability of the closed-loop system using the Generalized Nyquist Criterion.
27. The loop transmission is $T(s) = \frac{100}{s^2+s+10}$. There is a nonlinearity in the forward path. Determine a sector for which the system is absolutely stable using the Circle Criterion.
28. The loop transmission is $T(s) = \frac{100}{s^2-s+10}$. There is a nonlinearity in the forward path. Determine a sector for which the system is absolutely stable using the Circle Criterion.

Chapter 6
Feedback design – linear

Good control engineers should be paid by the dB.

– *Anonymous*

This chapter describes the design process, starting with the determination of the linear parts of the control system. After this design is complete, requisite nonlinear compensation is designed and implemented.

The goals of the design of the linear components are large feedback and sufficient relative stability in the nominal operating condition. As was determined in Chapter 3, characteristics of the system limit the feedback bandwidth. Given a maximum bandwidth, $\omega_b = 100$ rad/s, and functional bandwidth, $\omega_0 = 1$ rad/s, what should the loop transmission roll-off slope be for a well-designed feedback system?

Consider the loop transmission function frequency response approximation shown in Figure 6.1. The modulus slope is −6 dB/oct between the functional bandwidth and the 0 dB crossover frequency. The phase at crossover is approximately −90 degrees, and thus the phase margin is approximately 90 degrees. The gain margin is infinite.

Now consider the loop transmission function frequency response shown in Figure 6.2. The modulus slope is −12 dB/oct between the functional bandwidth and the 0 dB crossover frequency. The phase at crossover is slightly greater than −180 degrees, and thus the phase margin is very small. The gain margin, again, is infinite.

The second system has one more pole at the functional bandwidth frequency of 1 rad/s. With the steeper roll-off, the second system has approximately 40 dB more feedback over the functional bandwidth. However, it has very low phase stability margin, and thus excessive positive feedback in the neighborhood of crossover making the design useless. The first system will be stable. (Note: This system has no positive feedback at all, and the Bode sensitivity integral does not apply to this type of system. It is further noted that most practical systems will have a second-order or steeper roll-off at high frequency.) However, the phase stability margin is excessive (approximately 60 degrees greater than the Bode minimum). This excessive stability margin comes at the cost of reduced feedback over the functional bandwidth, reducing the control system's performance for no practical reason.

If the second-order roll-off is insufficiently stable in a relative sense and the first-order roll-off has insufficient feedback, then the superior design would have a roll-off slope between −6 and −12 dB/oct. From the phase/gain relationship for minimum phase systems described in Chapter 2, a modulus slope of −10 dB/oct has a phase of

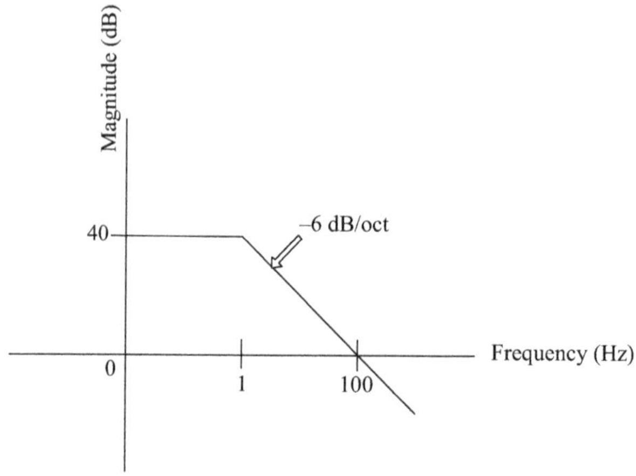

Figure 6.1 Loop transmission with a first-order roll-off

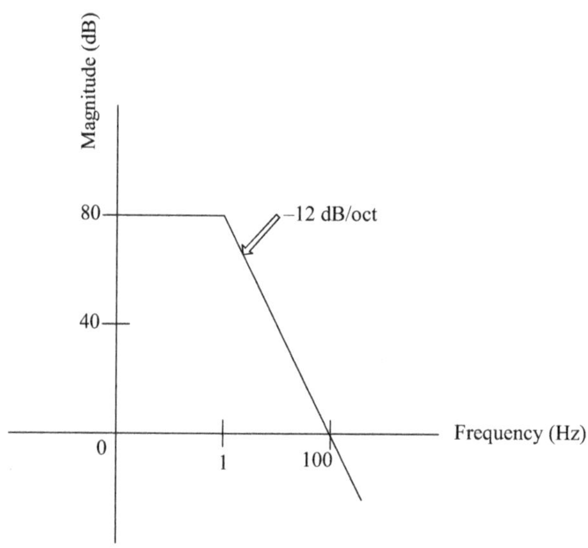

Figure 6.2 Loop transmission with a second-order roll-off

−150 degrees. This results in a 30-degree phase margin at crossover. Additionally, if the plant modulus varies (moderately), the phase margin will be retained. This third controller is a good compromise between the robustness of the first system and the aggressiveness of the second system. It does, however, raise a concern from the perspective of synthesis. Poles contribute phase in integer multiples of −90 degrees. The second system has one more pole at the break frequency than the first and thus has 90 degrees less phase at high frequency (in relation to the break frequency). With a linear compensator modeled by a rational function, how does one realize a modulus slope in between first and second order?

6.1 The Bode loop response

The material in this chapter focuses on two approaches to linear control design for high performance. The first is referred to as the Bode loop response, which is segregated over three disjoint frequency intervals. At low frequency, a cascade of carefully spaced poles and zeros causes a roll-off slope beyond the functional bandwidth that is a compromise between aggressive feedback application and relative stability. After crossover, a lead network called the *Bode step* is employed to compensate for the lag caused by the aggressive roll-off at frequencies higher than this lead. This high-order compensator gives large feedback over the functional bandwidth, with good stability properties with particular plant parameter variations.

6.1.1 Shaping the response below crossover: roll-off slope in between first and second order

A network pole reduces the loop shape slope by −6 dB/oct, two poles by −12 dB/oct and so on. A good control design requires a slope in between these. Consider the following complex function:

$$T_{des} = e^{\log(A_0)+\frac{5}{3}\log\theta(jf)} \tag{6.1}$$

where $\theta(jf) = \frac{1}{\sqrt{1-f^2+jf}}$, and f is the frequency normalized with respect to functional bandwidth. The modulus and argument of this function are shown in Figure 6.3. The modulus is flat to the functional bandwidth, then transitions to a roll-off of −10 dB/oct. This response has a phase of −150 degrees at all frequencies higher than the functional bandwidth, thus a 30-degree phase margin at whatever crossover frequency beyond the functional bandwidth. This is a desirable loop shape for a linear controller.

The synthesis of T_{des} is of concern, as it is not a rational function. A constant slope function can be decomposed into a product of rational and irrational functions $s^{-p} = s^{-m}s^{-q}$, where m is an integer and $0 < q < 1$. The modulus slope of s^{-q} can be approximated by a network function of appropriately spaced poles and zeros.

$$q = \frac{6b}{a+b} \tag{6.2}$$

where a and b are the octave spacing from zero to pole and pole to zero, respectively. For example, consider a nine-octave interval starting at 0.25 rad/s where the desired modulus slope is −10 dB/oct. A fourth-order filter is designed to have an approximately −4 dB/oct roll-off. A zero/pole spacing of one octave is chosen ($a = 1$), so the pole/zero spacing is two octaves. The filter transfer function is as follows:

$$T(s) = \frac{(s+1)(s+8)(s+64)}{(s+0.25)(s+2)(s+16)(s+128)} \tag{6.3}$$

Figure 6.4 shows the frequency response of this system. The roll-off is approximately −4 dB/oct. To realize the desired −10 dB/oct roll-off, the

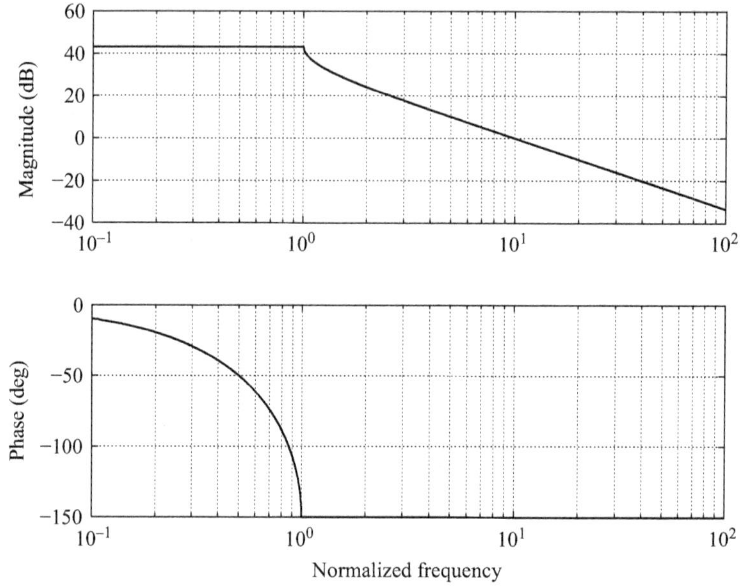

Figure 6.3 Optimal loop transmission shape at frequencies below the crossover

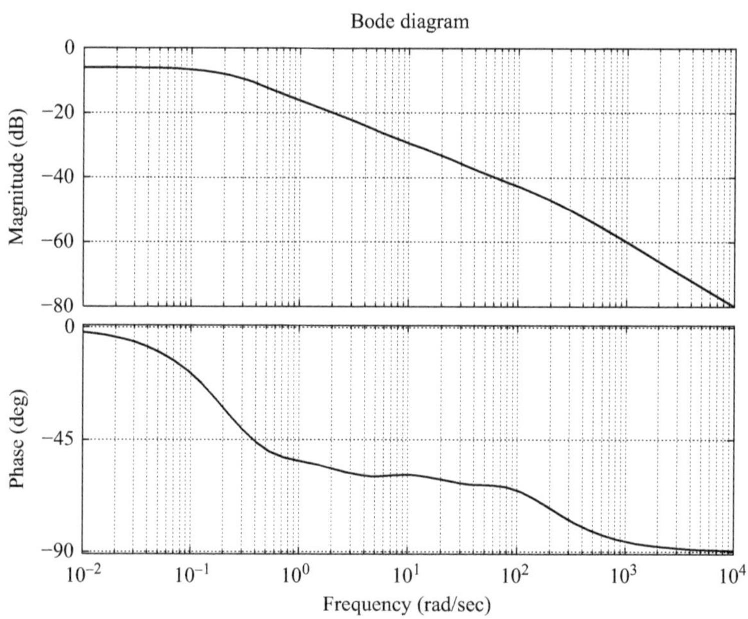

Figure 6.4 A system with pole zero spacing to approximate a −4 dB/oct roll-off

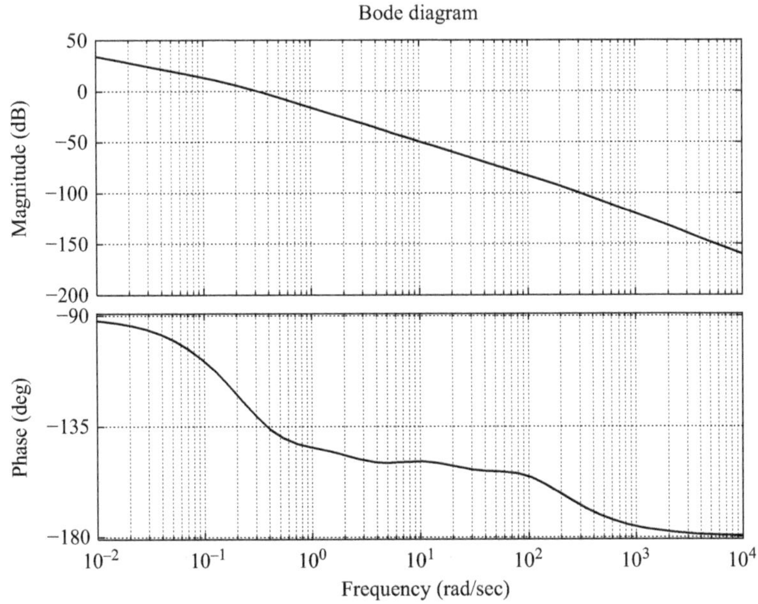

Figure 6.5 A system with pole zero spacing to approximate a −10 dB/oct roll-off

characteristic polynomial of $T(s)$ is augmented with an origin pole. Figure 6.5 shows the frequency response with the desired −10 dB/oct roll-off over the nine-octave interval.

6.1.2 *Shaping the loop response above crossover: high-frequency slope and the Bode step*

Plant parameter uncertainty and sensor noise tend to increase at high frequencies. To avoid unwanted control action at these frequencies, the feedback compensator should be designed so that feedback becomes negligible at as low a frequency beyond crossover as is practical. This requires a high-frequency loop transmission slope steeper than that between the functional bandwidth and 0 dB crossover. Typically, this slope is $-6n$ dB/oct, n being an integer of value 2 or greater. Figure 6.6 shows a partial loop transmission function with a 0 dB crossover frequency of f_b and a high-frequency roll-off break frequency of f_c. The modulus slope is $-12(1-y)$ dB/oct (for a −10 dB/oct roll off, $y = \frac{1}{6}$) below f_b and $-6n$ dB/oct above f_c.

The next design consideration is, given real number y and integer n, to shape the response between frequencies f_b and f_c. The modulus slope at low frequency, defined by y, is selected for the combined goal of feedback at low frequency and adequate phase stability margin. The increase in the steepness of the slope at frequencies higher than f_c has the deleterious effect of decreasing the phase at lower frequencies. As such, the transition between f_b and f_c should provide phase lead to counteract this.

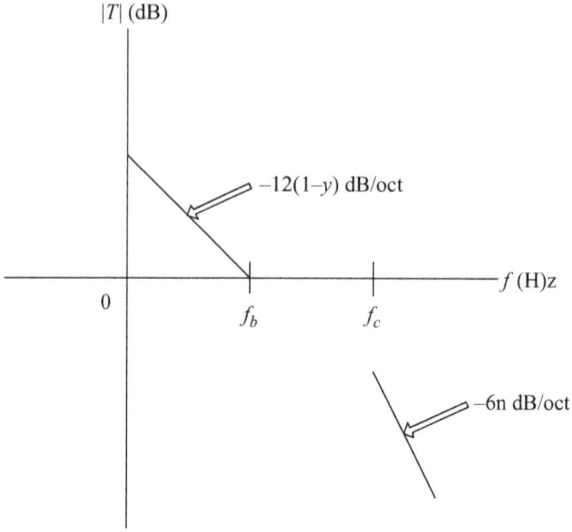

Figure 6.6 Partial loop transmission

Figure 6.7 shows the loop transmission with a lead transition between f_b and f_c. The slope of $-12(1-y)$ dB/oct is retained until f_d, where the modulus is $-x$ dB to frequency f_c. This lead is called the *Bode Step*, which compensates for the phase delay of the high-frequency roll-off and time delay.

Phase delay due to time delay increases linearly with frequency. The designer should not attempt to compensate for this with lead systems at frequencies where the phase delay due to time delay is large, as the loop gain increase will tend to

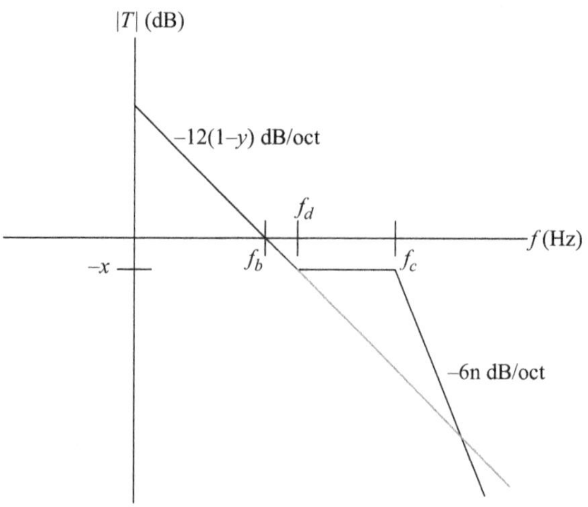

Figure 6.7 The Bode step

destabilize the system. A good rule of thumb is to limit the nonminimum phase delay at f_c to $|B_n(f_c)| < 1$ rad. For a time delay of t_d, the maximum f_c is approximately $0.16\frac{1}{t_d}$. The nonminimum phase lag at frequency $f < f_c$ can be approximated as follows:

$$B_n \approx \ |B_n(f_c)|\frac{f}{f_c} \tag{6.4}$$

Recall from the discussion of the Bode phase/gain relationship in Chapter 2 that the phase delay from n poles at f_c at $f < f_c$ can be approximated by $\frac{2}{\pi}\,n\frac{f}{f_c}$.

Consider the discarded ray in Figure 6.7. The approximate phase delay from this ray at $f < f_d$ is $\frac{2}{\pi}\,2(1-y)\frac{f}{f_d}$. This is the desired phase delay of the complete loop transmission at these frequencies (effectively equivalent to the loop shape with constant slope $-12(1-y)$ dB/oct at all frequencies). Equating this phase delay to the combined delay from nonminimum phase and the high-frequency roll-off gives the required Bode step width $\frac{f_c}{f_d}$.

$$\frac{2}{\pi}n\frac{f}{f_c} + |B_n(f_c)|\frac{f}{f_c} \approx \frac{2}{\pi}2(1-y)\frac{f}{f_d} \tag{6.5}$$

$$\frac{f_c}{f_d} \approx \frac{n + \frac{\pi}{2}|B_n(f_c)|}{2(1-y)} \tag{6.6}$$

Example: The crossover frequency is $f_b = 100$ Hz. The high-frequency roll-off slope is to be -18 dB/oct. The loop transmission slope at low frequency is to be -10 dB/oct. The Bode step gain is to be -6 dB. There is no nonminimum phase delay. Design the Bode step.

The low-frequency roll-off is -10 dB/oct, so $y = \frac{1}{6}$. The high-frequency roll-off is -18 dB/oct, so $n = 3$. For a Bode step gain of -6 dB, the -10 dB/oct roll-off is maintained to f_d, where $\log_2(\frac{f_d}{f_b}) = \frac{6}{10}$, or $f_d = 1.52 f_b = 152$ Hz. Since there is no nonminimum phase, the Bode step width is $\frac{f_c}{f_d} \approx \frac{3}{2(1-\frac{1}{6})} = 1.8$ and $f_c = 274$ Hz.

The Bode step is a lead and as such increases the loop gain at frequencies higher than crossover. This feature is a potential stability threat if there are unmodeled dynamics at frequencies in the neighborhood of the step. As such, the width should be designed only as wide as is necessary to compensate for the phase delay of the high-frequency slope.

6.1.3 The complete loop shape

Figure 6.8 shows the loop transmission designed for a particular functional bandwidth. A complex function similar to (6.1) is used to shape the loop response over the functional bandwidth. In doing so, the feedback over the functional bandwidth can be increased by the gain reduction over one octave in the roll-off. For example, if the loop transmission roll-off slope is -10 dB/oct beyond the functional bandwidth, the feedback over the functional bandwidth is increased by 10 dB using this method of loop shaping. Of course, a rational function with appropriately spaced poles must be found to approximate the frequency response of this complex function.

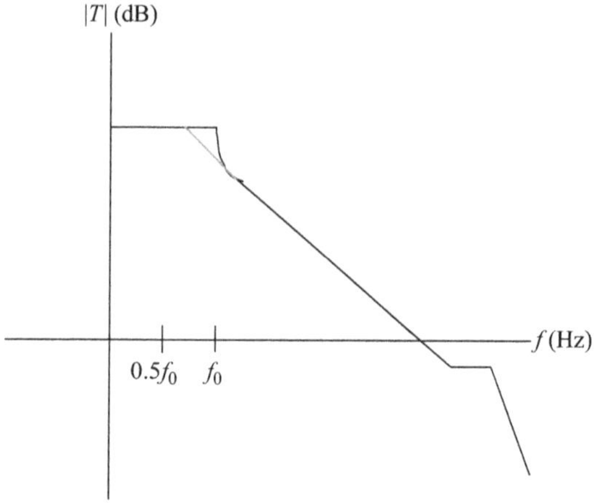

Figure 6.8 The complete loop response

It is evident that a well-shaped loop transmission requires a much higher order feedback compensator than more commonly implemented controllers. It is common for this type of compensator to be of order greater than 10, high when compared to a PID-type controller that, depending on additional high-frequency filtering, is typically of order 1–3. This disparity is of little concern, as increasing compensator order constitutes no significant increase in cost. A few more analog filters or a few more terms in a difference equation are insignificant compared to the increase in performance that a well-shaped compensator provides over common, low-order controllers. As such, a control designer should *always* endeavor to carefully shape the loop response to provide maximum feedback subject to bandwidth limitations.

6.1.4 Case study: loop shaping

Wind turbines are nonlinear systems, with characteristics that substantially restrict the available control system feedback: variable plant parameters, nonminimum phase zeros, adjacent pole pairs at low frequency, actuator delay and other limiting factors. Control performance, particularly in Region 3 (beyond rated wind speed), is critical, as the forces applied can severely reduce turbine lifetime.

The US Department of Energy suggests wind power capacity will increase by roughly an order of magnitude (305 GW) by 2030 [69]. To meet this goal, deployed turbines must be substantially larger than what are deployed today. In Reference 70, it is stated that 5 MW turbines or larger are needed to meet the goals of the 2030 report. These turbines will have hub heights of 140 and 153 m rotor diameters. Comparing this with the 80 m hub heights and 77 m rotor diameters of currently deployed 1.5 MW turbines prompts two questions related to automatic control applications for these plants: What control performance is required for these 5 MW turbines and what control performance is available?

For a constant torque wind turbine, Region 3 control has the primary goal of generator rate regulation. A low variance of this output is desired. Upper bounds on rate amplitude also apply subject to the limitations of the turbine power electronics. Ancillary Region 3 goals are related to load mitigation. This includes the use of variable blade pitch to damp tower modes and the use of generator torque to damp the shaft torsion mode.

Control performance is directly proportional to feedback, which in turn is a function of the available control bandwidth. Some plant characteristics, like flexible body dynamics and nonminimum phase, have the effect of reducing available bandwidth. It is expected that the flexible modes of proposed very large wind turbines (5 MW and larger) will be at lower frequencies than those of current turbines, and thus the available feedback is less. It is expected that the maximum blade angle rates will be lower than those of smaller turbines, also limiting the available control performance due to saturation. As a consequence, it is expected that the available control performance for the 5 MW turbines is less than that of current systems.

Control architecture

The goals of the control design are twofold. The primary goal is turbine shaft rate regulation in Region 3 (CART2 is a constant torque wind turbine, and for purposes of this comparison, the 5 MW wind turbine is assumed to be constant torque as well). Second, flexible body modes are targeted for feedback application for load mitigation. The control variables are collective blade pitch and generator torque. The outputs are high-speed shaft rate and tower fore/aft acceleration. These are variables in two control systems: a single-input, single-output controller (generator torque input and high-speed shaft rate output) and a single-input, two-output (SITO) controller (collective blade pitch input and high-speed shaft rate and tower fore/aft acceleration are outputs). Figure 6.9 shows the block diagram for this controller.

The generator and the high-speed shaft are collocated and the frequency responses for both the CART2 (Figure 6.10) and the 5 MW turbine consist of alternating zero/pole pairs that facilitate high-performance feedback design due to phase constraints. Low-order feedback compensation for both turbines results in more than 20 dB attenuation of shaft vibrations at the modal frequency. This control has the added benefit of reducing the modulus of the high-speed shaft response to collective blade pitch, making easier the task of gain stabilization of that mode in the shaft rate regulation design problem.

The design of the collective blade pitch controller is more challenging and is the focus of this comparative analysis. The plant is SITO; the two outputs are high-speed shaft rate for the Region 3 regulation problem (primary) and tower fore/aft acceleration for the vibration suppression problem (secondary). In contrast to the generator/high-speed shaft rate frequency response, the collective blade pitch responses present challenging design problems and it is this controller that distinguishes the two turbines from the perspective of available performance.

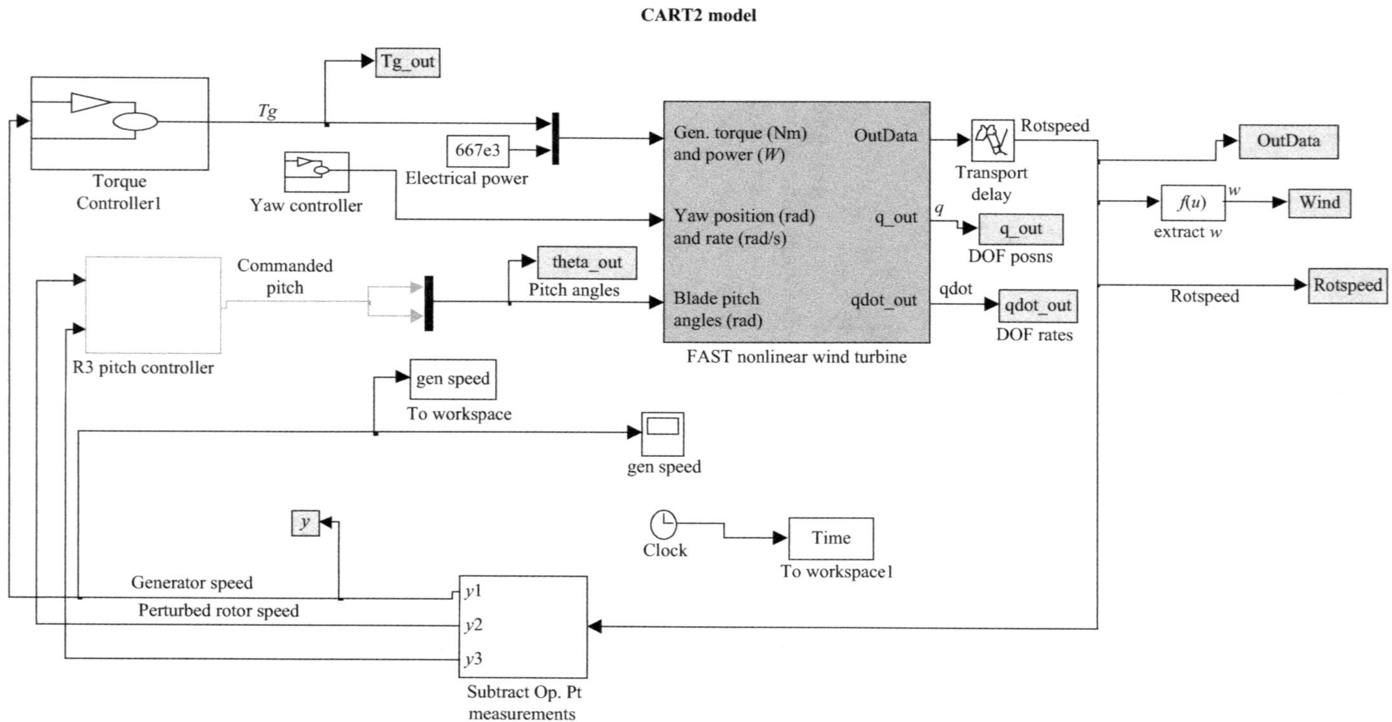

Figure 6.9   Region 3 control block diagram. Torque controller 1 is the SISO system, R3 pitch controller is the SITO system

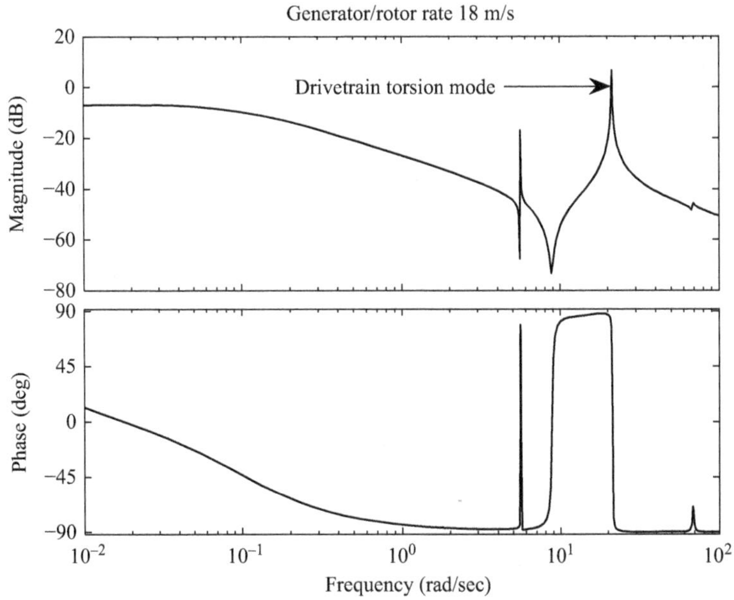

Figure 6.10 CART2 generator torque to high-speed shaft rate frequency response at 18 m/s

The comparative analysis is presented thusly:

1. Limitations to available feedback for the SITO collective blade pitch controller are identified and quantified for both the CART2 and 5 MW wind turbines.
2. The same loop transmission shape is applied to both the CART2 and 5 MW high-speed shaft rate control design that produces robust, high performance, linear control.
3. Using 1 and 2, the available feedback over a chosen functional bandwidth for both turbine control designs is calculated and compared.
4. Controller designs for both systems are developed and the performance is compared using the National Renewable Energy Laboratory's FAST (Fatigue Aerodynamics Structures and Turbulence) simulations of both wind turbines. These results are compared to the trend indicated in 3.

It is assumed that adequate driveshaft vibration attenuation is achieved via simple generator torque control, and details of this design are not included in the comparative analysis.

Comparison of available feedback: CART2 versus 5 MW

The performance difference between the collective blade pitch control performance of the 600 kW and 5 MW wind turbines is quantified. The following conditions apply to both systems:

1. The plant is two-output: high-speed shaft rate and tower fore/aft acceleration.
2. The control system is linear.

3. The blade pitch rate is limited. 20 deg/s for CART2 and 8 deg/s for the 5 MW turbine.
4. The functional bandwidth, defined as the frequency to which the loop transmission has a constant magnitude A_0, is 0.1 rad/s. This allows more than one decade for the loop transmission roll-off.
5. The loop transmission roll-off at frequencies higher than the functional bandwidth is -10 dB/oct. This approach combines aggressive feedback delivery with good robustness (-150 degree phase after the break).
6. If there are x octaves between the functional bandwidth and the 0 dB crossover frequency, then the loop transmission modulus over the functional bandwidth is $10(x+1)$ dB. This additional 10 dB of feedback is extracted by careful loop shaping at the functional bandwidth break and near 0 dB crossover.

The desired loop transmission function is given by equation 6.1 where $\theta(jf) = \frac{1}{\sqrt{1-f^2+jf}}$ and f is the frequency normalized with respect to functional bandwidth. Figure 6.11 shows the modulus and argument of this function. Implemented controllers will be defined by rational functions with poles and zeros selected to approximate this shape.

Available feedback for the CART2 Region 3 high-speed shaft rate regulator
Figures 6.12 and 6.13 show the CART2 high-speed shaft rate and tower fore/aft acceleration responses to collective blade pitch at 18 m/s wind speed, found using

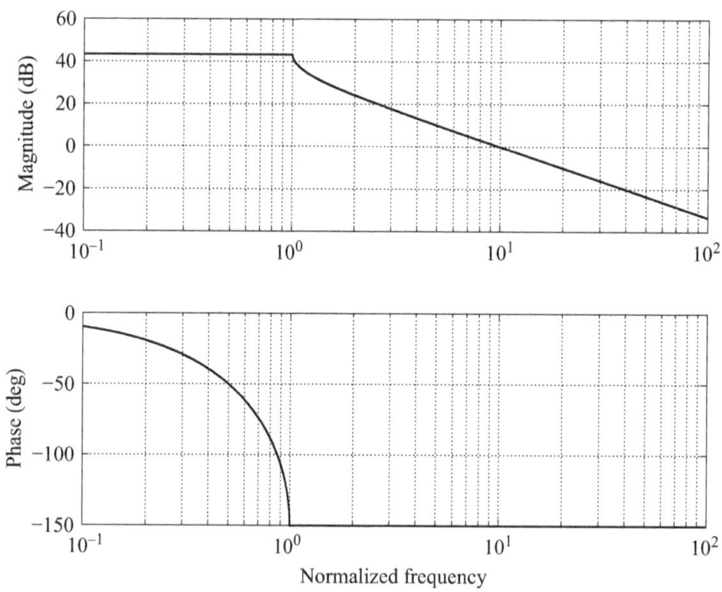

Figure 6.11 Desired loop transmission as a function of frequency normalized with respect to functional bandwidth

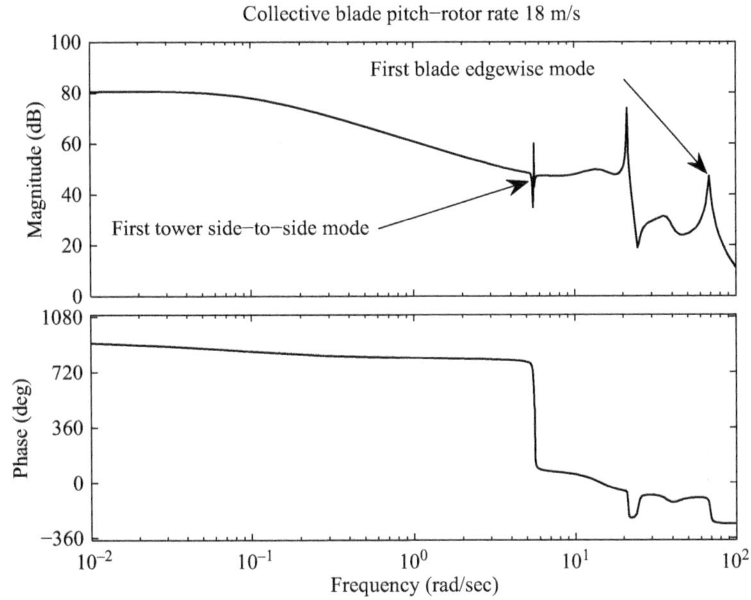

Figure 6.12 CART 2 collective blade pitch to high-speed shaft rate frequency response at 18 m/s

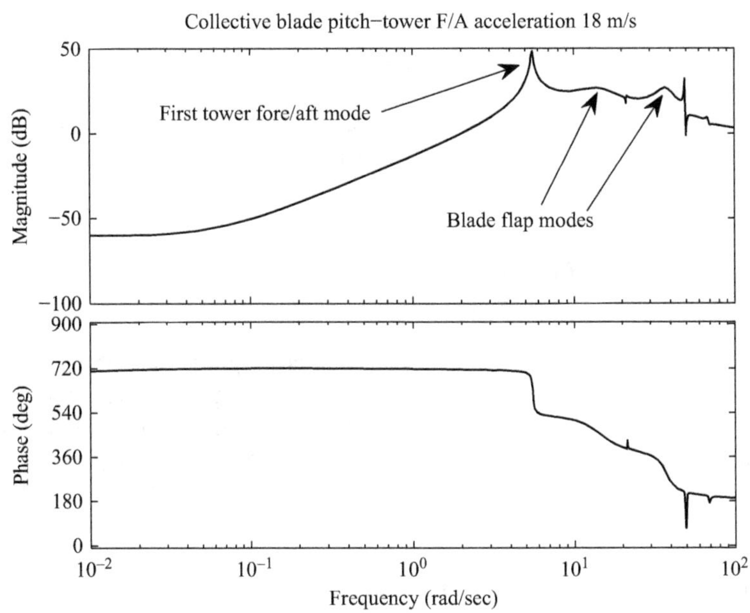

Figure 6.13 CART 2 collective blade pitch to tower fore/aft acceleration frequency response at 18 m/s

the FAST model of this turbine. These plots are used to determine the feedback limitations for the high-speed shaft rate controller, the primary controller for this application.

1. There is substantial nonminimum phase in the neighborhood of the first tower mode at 5.5 rad/s of the shaft rate response (Figure 6.12). This is the result of right-half plane zeros clustered near the first tower fore/aft and side-to-side modes at this frequency. The nonminimum phase delay at this frequency makes impossible the design of a shaft rate controller of higher bandwidth.
2. There must be at least one octave of separation between the frequencies where the high-speed shaft rate is negative and where the acceleration feedback is negative.
3. The first blade edgewise mode at approximately 70 rad/s in the shaft rate response must be either gain or phase stabilized. The drivetrain mode at 22 rad/s is reduced by the generator controller; however, the collective blade pitch controller will stabilize this mode in case the generator loop is opened.

Items 1 and 2 are the most substantial limitations. The nonminimum phase at 5.5 rad/s described in item 1 requires gain stabilization of modes near this frequency. This limits the bandwidth of the high-speed shaft rate controller to ~1–2 rad/s. Control of the first tower fore/aft mode (5.5 rad/s) requires acceleration feedback over the interval [4 6] rad/s, thus the restriction of item 2 limits the high-speed rate feedback loop to 2 rad/s. The bandwidth limit caused by item 3 is higher than those of 1 and 2 and is thus not a critical limitation.

The crossover frequency is ~1.5 rad/s. The interval from the functional bandwidth (0.1 rad/s) to the crossover frequency of 1.5 rad/s is 3.9 octaves wide. Assuming careful loop shaping at the break, the high-speed shaft controller can deliver $10(3.9 + 1) = 49$ dB of feedback over the functional bandwidth.

Available feedback for the 5 MW Region 3 high-speed shaft rate regulator
Figures 6.14 and 6.15 show the 5 MW high-speed shaft rate and tower fore/aft acceleration responses to collective blade pitch at 18 m/s wind speed. These plots are used to list the feedback limitations for the high-speed shaft rate controller.

1. There is substantial nonminimum phase in the neighborhood of the first tower mode at 2 rad/s of the shaft rate response (Figure 6.14). This is the result of right-half plane zeros clustered near the first tower fore/aft and side-to-side modes at this frequency. The nonminimum phase delay at this frequency makes impossible the design of a shaft rate controller of higher bandwidth.
2. There must be at least one octave of separation between the frequencies where the high-speed shaft rate is negative.
3. The first blade edgewise mode at approximately 25 rad/s in the shaft rate response must be stabilized (either gain or phase). The drivetrain mode at 10 rad/s is attenuated by the generator controller; however, the collective blade pitch controller will stabilize this mode in case the generator loop is opened.

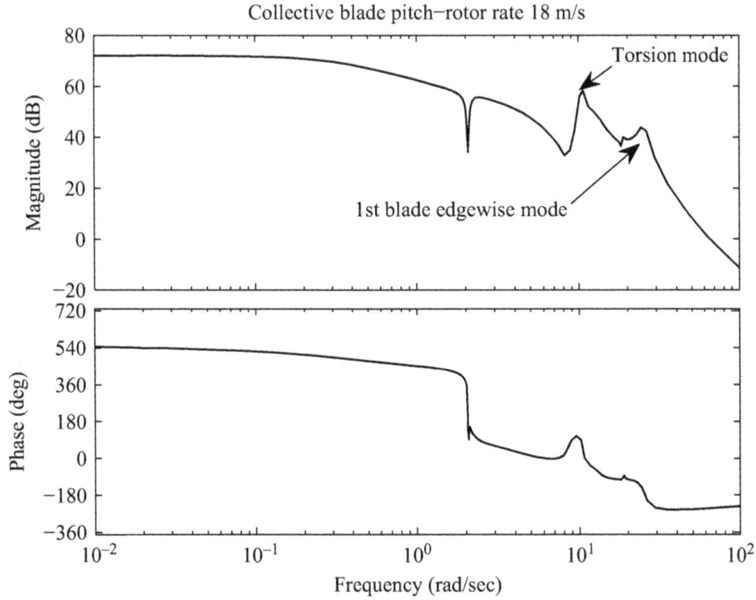

Figure 6.14 5 MW collective blade pitch to high-speed shaft rate frequency response at 18 m/s

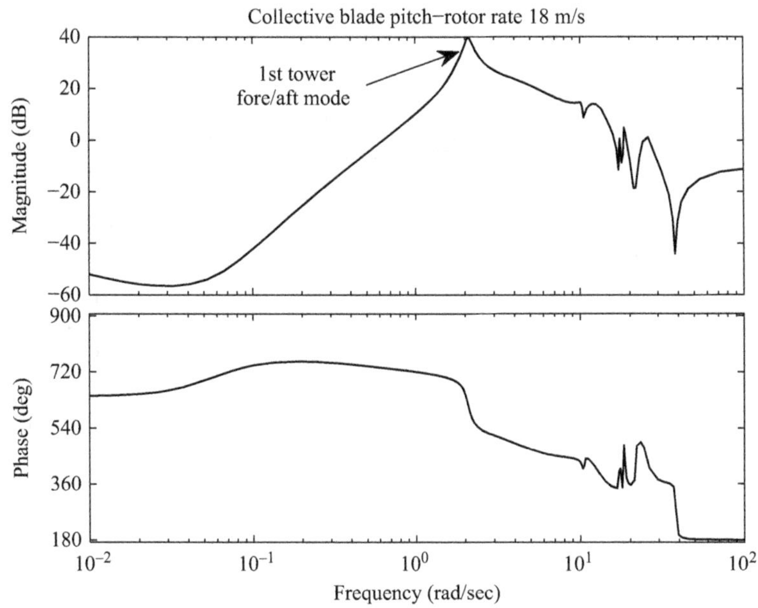

Figure 6.15 5 MW collective blade pitch to tower fore/aft acceleration frequency response at 18 m/s

As is the case for CART2, items 1 and 2 are the most substantial limitations. While this response does not have the sharp poles that CART2 has around 2 rad/s, the nonminimum phase at this frequency described in item 1 requires gain stabilization of modes near this frequency. This limits the bandwidth of the high-speed shaft rate controller to ~0.5–1 rad/s. Control of the first tower fore/aft mode (2 rad/s) requires acceleration feedback over the interval ~[1 4] rad/s, thus the restriction of item 2 limits the high-speed rate feedback loop to 0.5 rad/s. The driveshaft and edgewise modal frequencies are high enough so that item 3 is not a limitation.

The crossover frequency is ~0.5 rad/s. The interval from the functional bandwidth (0.1 rad/s) to the crossover frequency of 2 rad/s is 2.3 octaves wide. Assuming careful loop shaping at the break, the high-speed shaft controller can deliver $10(2.3 + 1) = 33$ dB of feedback over the functional bandwidth.

The more substantial limitations of the 5 MW turbine result in a factor of approximately six times less rate regulation performance than the CART2, theoretically.

6.2 Phase stabilization

A given linear, time invariant single-input, single-output (LTI SISO) frequency response of a stable plant is shown in Figure 6.16. The −180 degree phase contribution of the mode at 100 rad/s represents a potential feedback stability risk. The

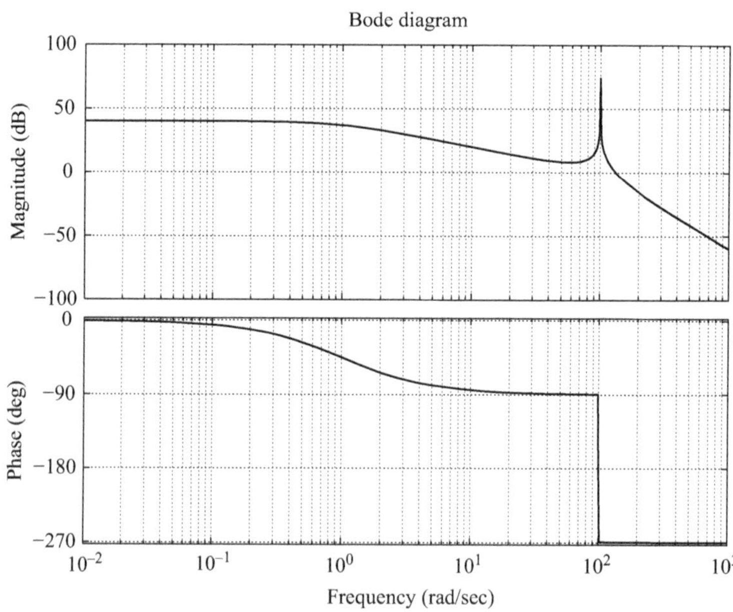

Figure 6.16 A plant frequency response with a lightly damped flexible mode

control design could reduce loop gain at the modal frequency (T-plane plot is confined within the unit circle). This is *gain stabilization*. Low-frequency poles or notching can accomplish this. Applying a very deep notch at the modal frequency reduces the gain; however, it is noted that the notch approach is dangerous for high-quality factor modes with potentially variable natural frequencies (Chapter 2). Low-pass filtering is a more robust method to gain stabilize the mode, but the quality factor of the mode is so high that the requisite gain reduction severely reduces the amount of feedback we can apply at low frequency.

An alternative to gain stabilization is *phase stabilization*. As opposed to reducing the gain at the modal frequency to avoid the critical point, the phase is reduced to rotate the T-plane plot away from the critical point, allowing greater than unity gain at the modal frequency. The plant Nyquist plot is shown in Figure 6.17. Note the large loops in the left half plane associated with the flexible mode that encircle the critical point. The compensator $C(s) = \frac{1000}{s^2+210s+2000}$ is designed for the feedback system. The loop transmission function Nyquist plot is shown in Figure 6.18. The compensator phase delay rotates the loops away from the critical point, and the closed-loop system is stable despite the gain at the modal frequency being greater than one.

Phase stabilization is typically applied to modes beyond the initial 0 dB crossover, at frequencies where plant knowledge is less than perfect. As such, phase stabilization should only be implemented at frequencies where there is decent plant phase knowledge and low phase variation. If there is unexpected plant phase advance in the neighborhood of the modal frequency, the loop in the T-plane plot associated with the mode may rotate counterclockwise and encircle the critical point.

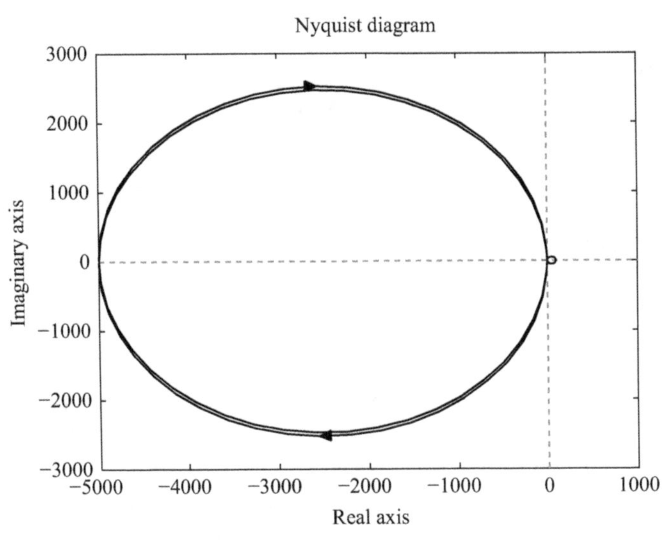

Figure 6.17 Nyquist plot of plant with flexible modes

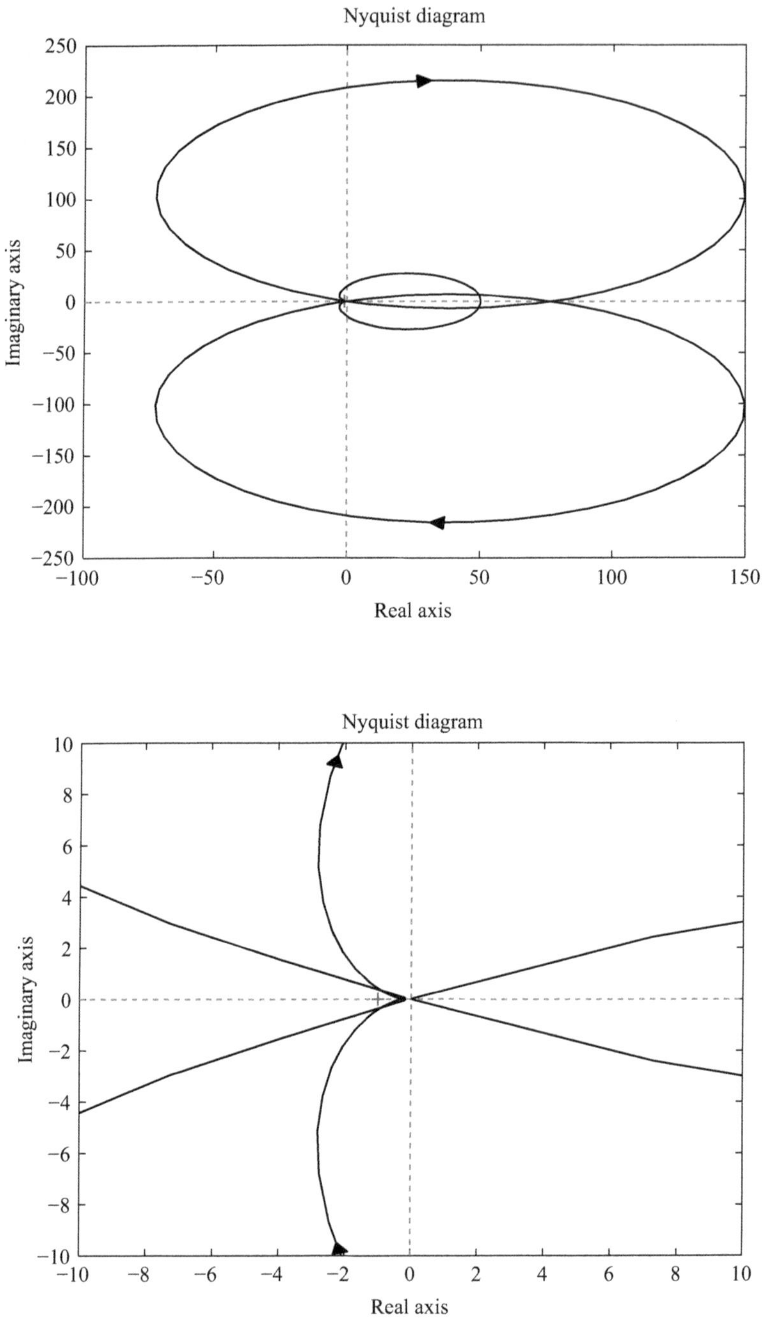

Figure 6.18 T-plane plot of a phase-stabilized system

6.3 Nyquist-stable system

The loop transmission shape described in Section 6.1 is a combination of good performance and robustness that make it an effective design for many LTI SISO applications. For applications where very large feedback is required, an alternative to this loop shape is required. Figure 6.19 shows the T-plane plot for a stable loop transmission function. Note that the plot intersects the negative real axis outside the unit circle, yet does not encircle the critical point. This is an example of a *Nyquist-stable* system. This is not to be confused with a system that simply satisfies the Nyquist Stability Criterion, which is not sufficient for Nyquist stability. The Nyquist-stable system has a roll-off greater than -12 dB/oct, then using lead compensation 'unwraps' the Nyquist locus to avoid encircling the critical point. This very steep roll-off results in more feedback over the functional bandwidth of the control system.

The Nyquist-stable system, of course, has trade-offs. From the Bode sensitivity integral, it is clear that this controller will have greater positive feedback that must be carefully distributed over frequencies where disturbance power is low. A more severe trade-off is the conditional nature of this system's stability. The T-plane plot in Figure 6.19 clearly indicates the system does not satisfy the Circle Criterion for sector [0 1]. In fact, the Nyquist-stable controller has a tendency for oscillation when there is a saturation in the loop. The effective drop in loop gain caused by the saturation reduces the crossover frequency, and the loop outside the unit circle in Figure 6.19 collapses toward the origin, encircling the critical point.

6.4 Two-input, single-output control

Some control applications require dynamic ranges beyond which a single actuator can deliver. For instance, consider a position regulator problem where open-loop output amplitude driven by disturbances is on the order of centimeters, and the closed-loop output must be on the order of nanometers. To control to such small amplitudes, the actuator must be very accurate, and as extremely large feedback is required to achieve this level of disturbance rejection, the actuator must also be capable of very high bandwidths. So-called 'smart' actuators, such as those using piezoelectric or electrostrictive materials, have these qualities. However, these actuators are typically capable of tens or hundreds of micrometers of motion, clearly insufficient to reject disturbances generating centimeters of motion. An alternative actuator, such as a voice coil, that has the requisite stroke typically will not have the accuracy or bandwidth to satisfy the requirements of the problem.

The solution is to employ both actuators in a two-input, single-output (TISO), configuration. The block diagram of such a system is shown in Figure 6.20. The plant and compensator transfer matrices are $P(s) = [p_1(s)\ p_2(s)]$ and $C(s) = [c_1(s)\ c_2(s)]^T$, respectively. The fine motion actuator can be considered the primary actuator ($p_2(s)$), providing the accuracy required. The coarse actuator ($p_1(s)$) serves

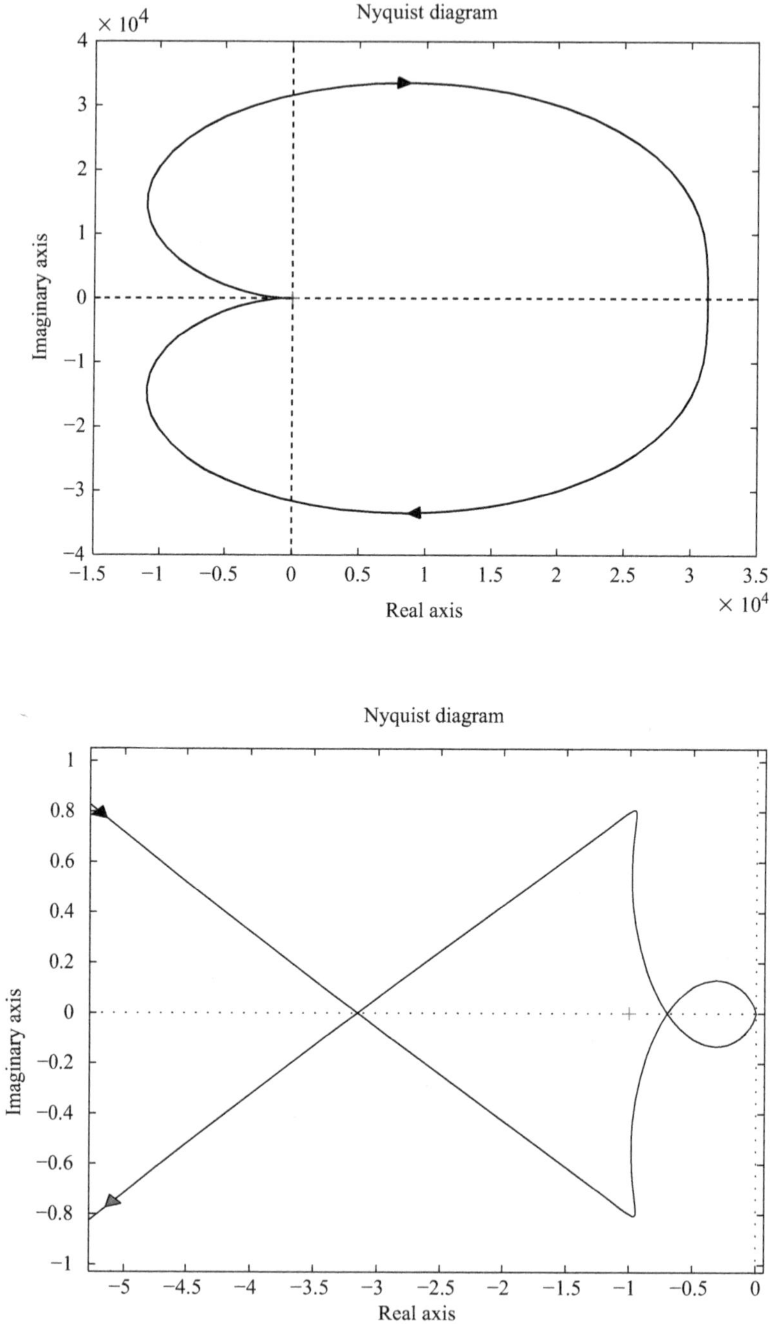

Figure 6.19 T-plane plot of a Nyquist-stable system

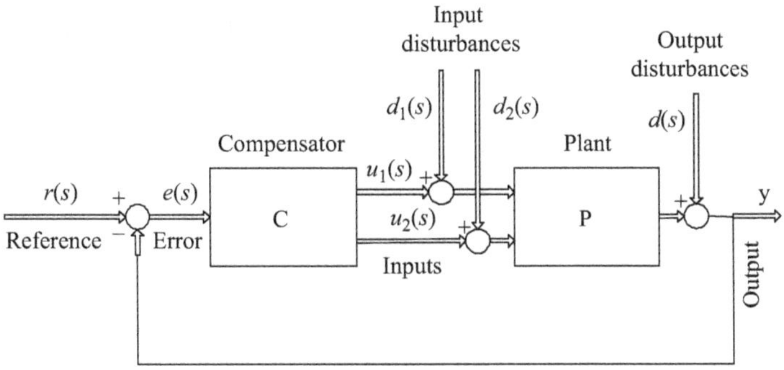

Figure 6.20 TISO system block diagram

to keep the fine motion actuator from saturating due to the large amplitude disturbances at low frequency. Alternative compensator architectures include a cascade connection of $c_1(s)$ and $c_2(s)$ at the input of coarse actuator.

Consider the response of the TISO system to reference input, $r(s)$, disturbances at the input of the plant $D(s) = [d_1(s)\ d_2(s)]^T$ and disturbances at the output of the plant $d(s)$.

$$y_u(s) = \frac{P(s)C(s)}{1+P(s)C(s)}r(s) = \frac{p_1(s)c_1(s)+p_2(s)c_2(s)}{1+p_1(s)c_1(s)+p_2(s)c_2(s)}r(s) \tag{6.8}$$

$$\begin{aligned} y_D(s) &= \frac{P(s)}{1+P(s)C(s)}D(s) \\ &= \frac{p_1(s)}{1+p_1(s)c_1(s)+p_2(s)c_2(s)}d_1(s) \\ &\quad + \frac{p_2(s)}{1+p_1(s)c_1(s)+p_2(s)c_2(s)}d_2(s) \end{aligned} \tag{6.9}$$

$$y_d(s) = \frac{1}{1+P(s)C(s)}d(s) = \frac{1}{1+p_1(s)c_1(s)+p_2(s)c_2(s)}d(s) \tag{6.10}$$

The loop transmission function is $T(s) = p_1(s)c_1(s) + p_2(s)c_2(s) = T_1(s) + T_2(s)$, a summation of two-loop transmission functions that are shaped for different tasks. For instance, the coarse compensator $c_1(s)$ is designed for large feedback at low frequency, and the fine motion compensator $c_2(s)$ provides feedback at high frequency. This approach prevents the fine motion actuator from saturation due to large disturbances at low frequency and provides high accuracy at high frequency.

It is interesting to note that in a coarse/fine actuation control strategy, only $T_2(s)$ needs to be carefully shaped near unit gain for stability. This is because the coarse actuator feedback becomes negligible at a frequency where there is still substantial feedback for the fine actuator (i.e. $T(s) \approx c_2(s)p_2(s)$). As such, only $c_2(s)$ must be carefully shaped for adequate relative stability. However, if $T_1(s)$ does not satisfy the Nyquist Criterion, then the system stability is threatened if there is a reduction in the loop gain $|T_2(s)|$ (i.e. $T(s) \approx c_1(s)p_1(s)$). This can occur if the fine actuator saturates, or the fine actuator loop is opened. Another consideration is the relative phase in the neighborhood of the "handoff" frequency ω_h where $|T_1(j\omega_h)| = |T_2(j\omega_h)|$. If these functions are out of phase, there is a zero at ω_h, which can be problematic. If they are in phase, there is a local boost in modulus that may not be desirable.

Example: TISO controller

Consider a TISO controller with the following loop transmission functions:

$$T_1(s) = \frac{10^6}{s^2(s+10)} \tag{6.11}$$

$$T_2(s) = \frac{10^4(s+1000)}{s(s+100)} \tag{6.12}$$

This is typical for a coarse and fine actuator system, where the coarse loop transmission has large gain at low frequency ($T_1(s)$) to desaturate the fine actuator, whose loop transmission ($T_2(s)$) has a high crossover frequency. Figure 6.21 shows the frequency response of these loop transmission functions, and the TISO loop transmission (summation). Note that careful loop shaping for stability is necessary only for $T_2(s)$. Figure 6.22 shows the step response of the closed-loop system.

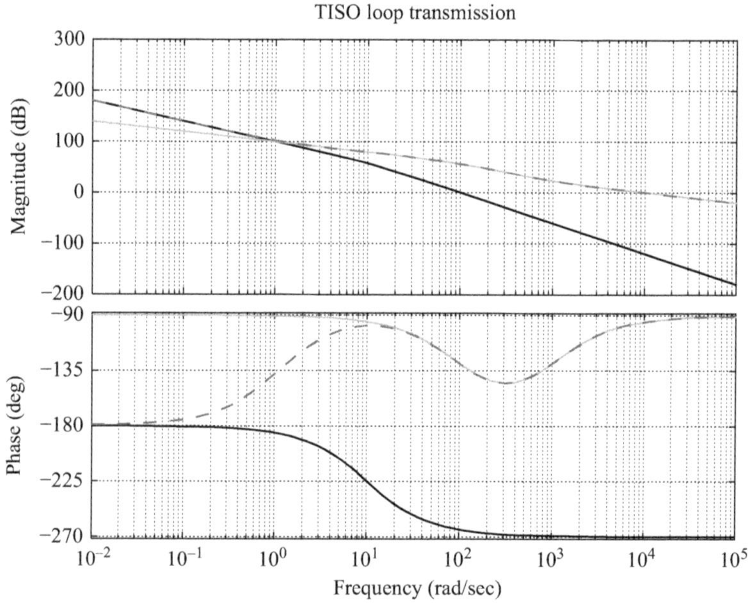

Figure 6.21 Coarse (solid black lines), fine (grey lines) and TISO (dotted lines) loop transmissions

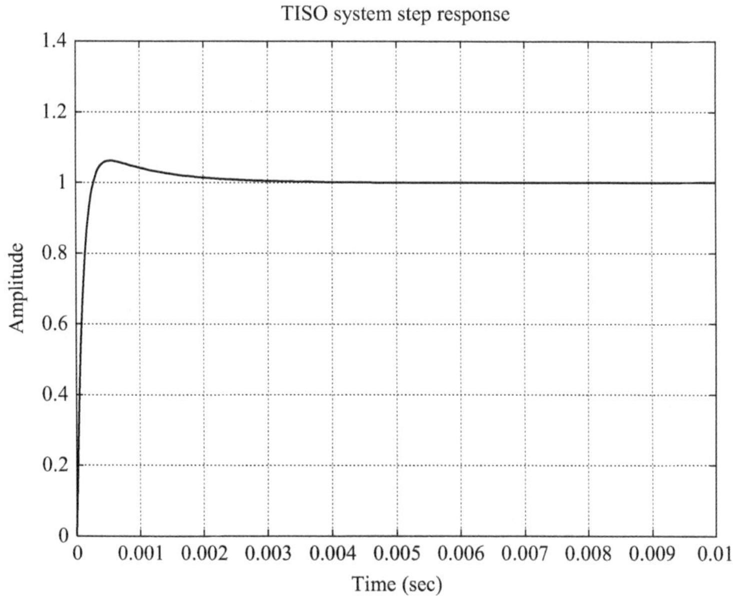

Figure 6.22 Step response of TISO system

6.5 Single-input, two-output control

In the previous section, we consider control designs where two actuators are used to control one output. Here, we consider the converse (Figure 6.23). These design problems arise when a single actuator drives two outputs, and it is either not practical or impossible to introduce a second actuator. $P(s) \in \mathbf{F}^{2\times 1}(s)$, $P(s) = [p_1(s)\ p_2(s)]^T$ describes the plant and $C(s) \in \mathbf{F}^{1\times 2}(s)$, $C(s) = [c_1(s)\ c_2(s)]$ the compensator. Although the system involves a plant with two outputs, the

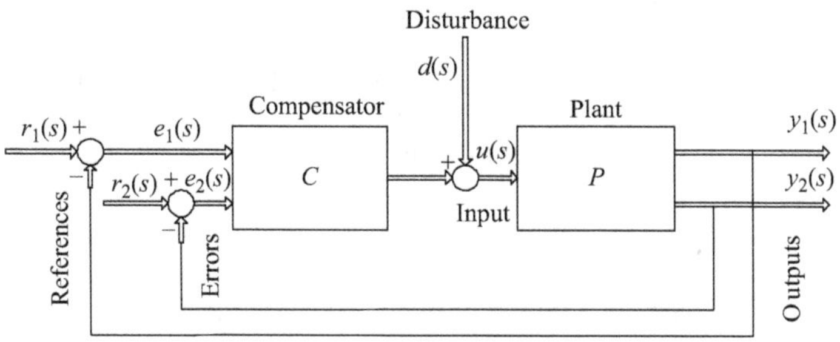

Figure 6.23 SITO system block diagram

Bode/Nyquist methods of stability analysis and performance assessment for SISO systems can be used for SITO designs. This is because the loop can be opened at the actuator and the return ratio found at this point, $T(s) = C(s)P(s) = c_1(s)p_1(s) + c_2(s)p_2(s)$. Using this function, the Nyquist Stability Criterion can be used to assess the stability of the SITO closed-loop system.

The reference tracking of the SITO controller is now considered.

$$Y(s) = (I + P(s)C(s))^{-1}P(s)C(s)R(s) \tag{6.13}$$

It is noted that matrix $P(s)C(s)$ is rank deficient, so some reference vectors $R(s) = [r_1(s)\ r_2(s)]^T$ cannot be tracked. Often the control goal is mixed tracking/regulation, and this is not an issue. Consider the rejection of disturbances at the input to the plant. The response $Y(s)$ to disturbance at the plant input, $d(s)$, is

$$Y(s) = (I + P(s)C(s))^{-1}P(s)d(s) \tag{6.14}$$

$$= \begin{bmatrix} 1 + p_1(s)c_1(s) & p_1(s)c_2(s) \\ p_2(s)c_1(s) & 1 + p_2(s)c_2(s) \end{bmatrix}^{-1} \begin{bmatrix} p_1(s) \\ p_2(s) \end{bmatrix} d(s) \tag{6.15}$$

$$= \begin{bmatrix} \dfrac{(1 + p_2(s)c_2(s))p_1(s) - p_1(s)p_2(s)c_2(s)}{(1 + p_1(s)c_1(s))(1 + p_2(s)c_2(s)) - p_1(s)p_2(s)c_1(s)c_2(s)} \\ \dfrac{(1 + p_1(s)c_1(s))p_2(s) - p_1(s)p_2(s)c_1(s)}{(1 + p_1(s)c_1(s))(1 + p_2(s)c_2(s)) - p_1(s)p_2(s)c_1(s)c_2(s)} \end{bmatrix} d(s) \tag{6.16}$$

The rational functions $p_1(s)c_2(s)$ and $p_2(s)c_1(s)$ cross-couple the disturbance at frequencies where the modulus is large. Consider the condition that the functions $c_1(s)p_1(s)$ and $c_2(s)p_2(s)$ have large moduli over disjoint frequencies. The condition is an intuitive separation of frequencies where compensator and the plant not associated with that compensator gains are large. If this condition is satisfied, the response to the input disturbance is

$$Y(s) \simeq \begin{bmatrix} \dfrac{p_1(s)}{1 + p_1(s)c_1(s)} \\ \dfrac{p_2(s)}{1 + p_2(s)c_2(s)} \end{bmatrix} d(s) \tag{6.17}$$

The responses are nearly decoupled and are inversely proportional to the feedback in each channel as two independent SISO systems would be. In contrast to the TISO system, where overlap of individual loop transmissions is acceptable, the concept of *frequency separation* is of paramount importance in SITO control. This involves the design of two compensators for the two different error signals that deliver negative feedback over frequency intervals that are disjoint. Failure to design adequate frequency separation in a SITO control system results in an input to the actuator that is a combination of corrective signals of two disparate filtered errors. This is related to the concept of controller/plant *alignment*, which can have a significant effect on output sensitivity.

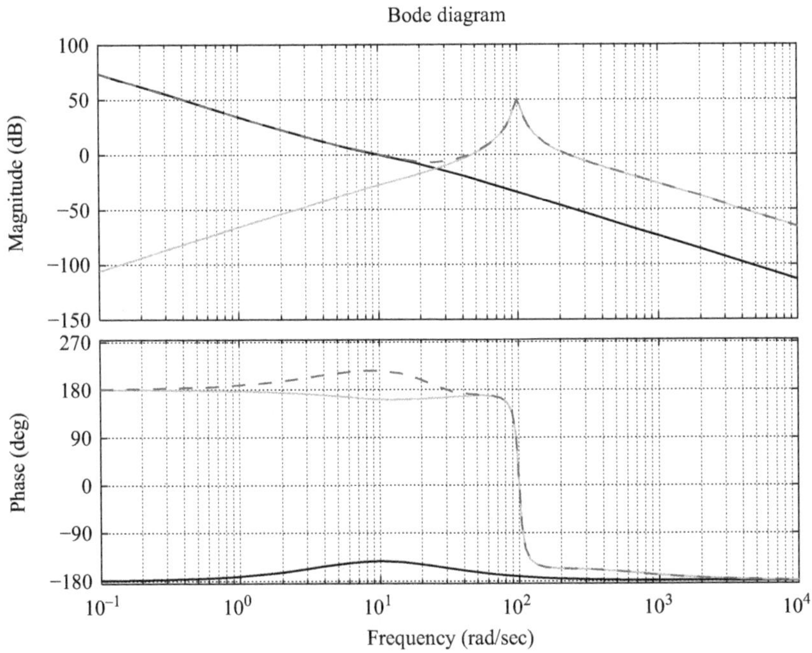

Figure 6.24 Frequency responses of $C_1(s)P_1(s)$, $C_2(s)P_2(s)$ and the SITO loop transmission

The intuitive notion that SITO systems are more restricted in available performance than TISO systems is supported in this analysis. Frequency separation reduces feedback, as a single actuator must perform a dual task. As such, SITO feedback systems trade performance for reduced cost.

Example: SITO control

Consider a SITO plant with a transfer function $P(s) = \left[\frac{100}{s^2} \; \frac{1000s^2}{s^3(s^2+10s+10{,}000)}\right]^T$. The feedback compensator is $C(s) = \left[\frac{2(s+5)}{s+20} \; 0.5\frac{(s+20)(s+200)}{(s+10)(s+400)}\right]$. Figure 6.24 shows the frequency responses of $c_1(s)p_1(s)$, $c_2(s)p_2(s)$ and the SITO loop transmission. Note the frequency separation of loop gains. The step response of the closed-loop system is shown in Figure 6.25.

6.5.1 *Case study: SITO control of a ship's rudder*

A ship's rudder couples to both yaw and roll motion, as depicted in Figure 6.26. In this application, the rudder serves as the actuator in a two-output closed-loop system with a block diagram shown in Figure 6.27. The rudder to yaw/roll

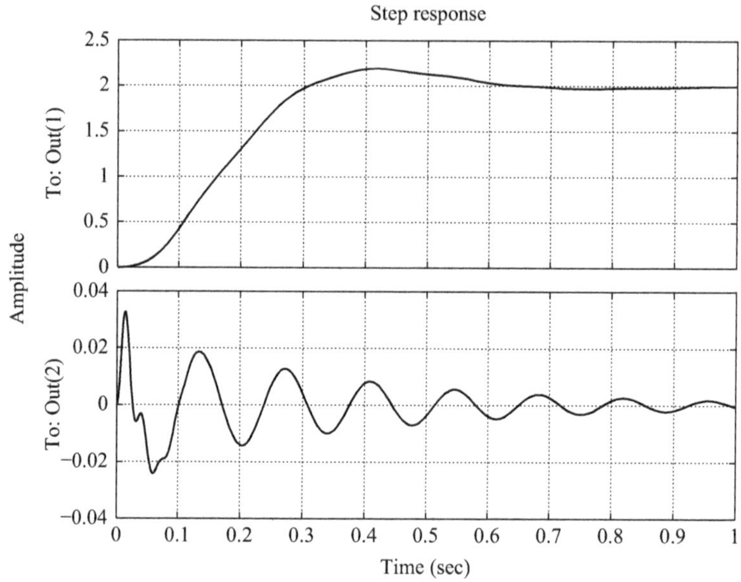

Figure 6.25 SITO closed-loop step response

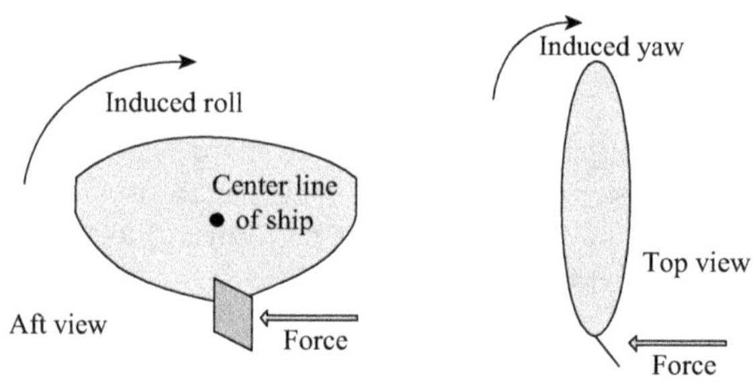

Figure 6.26 Roll and yaw moments generated by rudder deflection

plant model is taken from Van der Klugt [76]. Transfer functions from rudder angle to roll and yaw angles, $G_r(s)$ and $G_y(s)$, respectively, are given below

$$G_r(s) = \frac{0.3381s - 0.0169}{10s^3 + 2.18s^2 + 4.087s + 0.3969} \tag{6.18}$$

$$G_y(s) = \frac{0.2106s + 0.05694}{s(16.67s^2 + 11.67s + 1)} \tag{6.19}$$

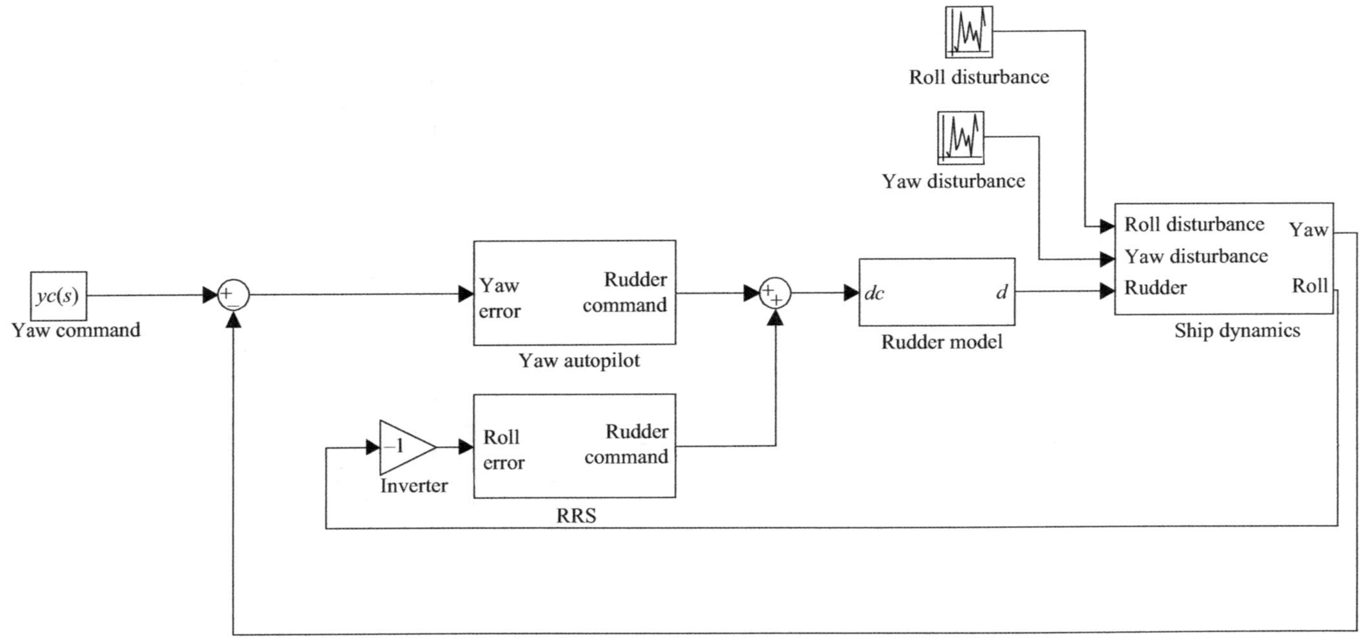

Figure 6.27 The combined heading and rudder roll stabilization (RRS) control system

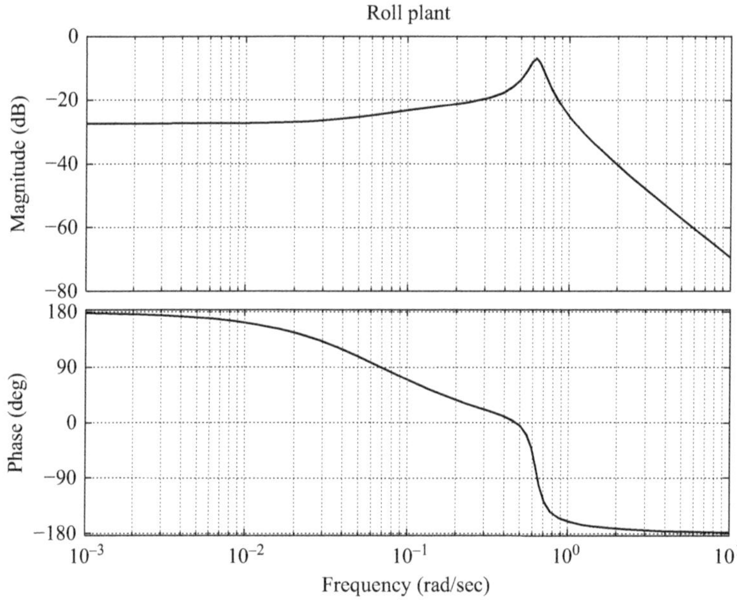

Figure 6.28 Roll plant frequency response

The roll plant frequency response is shown in Figure 6.28. The right-half plane zero at 0.05 rad/s has the capacity to complicate the roll controller design and to deliver large closed-loop sensitivity at low frequency. As the roll and yaw plants are coupled, there must be sufficient frequency separation in the individual control loops.

Rudder model

The simplified mathematical model of the rudder control loop of Van Amerongen [72] is used in this paper. A diagram of this system is shown in Figure 6.29, and a good description of the steering machine dynamics is given in Reference 75.

The rudder is limited in angle, and the hydraulic steering machine is limited in rate, the effects of which are modeled as saturations (rudder limiter and rudder rate limiter, respectively) in Figure 6.29. These saturations limit performance and potentially threaten the stability of the feedback system. The angle limit in this paper is 35 degrees. An interval of rudder rate limits is used in the subsequent analysis.

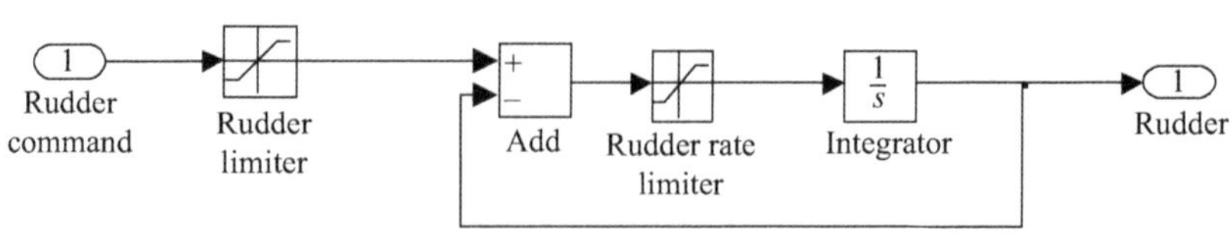

Figure 6.29 Simplified rudder model with angle and rate limits

Rudder roll stabilization controller

Three principal design issues are considered for the new roll stabilization controller. First, the wave disturbance spectrum is concentrated in the decade from 0.1 to 1 rad/s. This, plus the fact that the actuator is not very effective in frequencies higher than 1 rad/s, suggests that the maximum available feedback should be applied in this interval. Second, the coupled yaw and roll plants require frequency separation between the heading and roll stabilization controllers. The new roll controller will be designed to cross 0 dB at no less than 0.2 rad/s, which is the best case scenario. The third consideration is the nonminimum phase zero in the roll plant. It is fortunate that this zero is two octaves lower in frequency than the minimum first crossover frequency, as its phase contribution is only about 105 degrees at 0.2 rad/s.

An 8th-order roll stabilizing controller is designed with these three issues taken into consideration. The gain, zeros, and poles for the compensator C_r are $K = 79433$, $s_z = (0, -0.6000 \pm 1.3748i, -0.1800 \pm 0.2400i, -1.200, -0.5000)$ and $s_p = (-0.05, -2.400 \pm 3.624i, -0.6000 \pm 0.8000i, -0.0050 \pm 0.7000i, -100)$. The low-frequency poles and zeros are spaced for a more aggressive roll-up/roll-off than is available with low-order compensation. A lead is applied to boost the phase at the second crossover. The simple pole at 100 rad/s reduces loop gain at high frequency and provides a strictly proper compensator transfer function. Figure 6.30 shows the return ratio of the 8th-order roll controller.

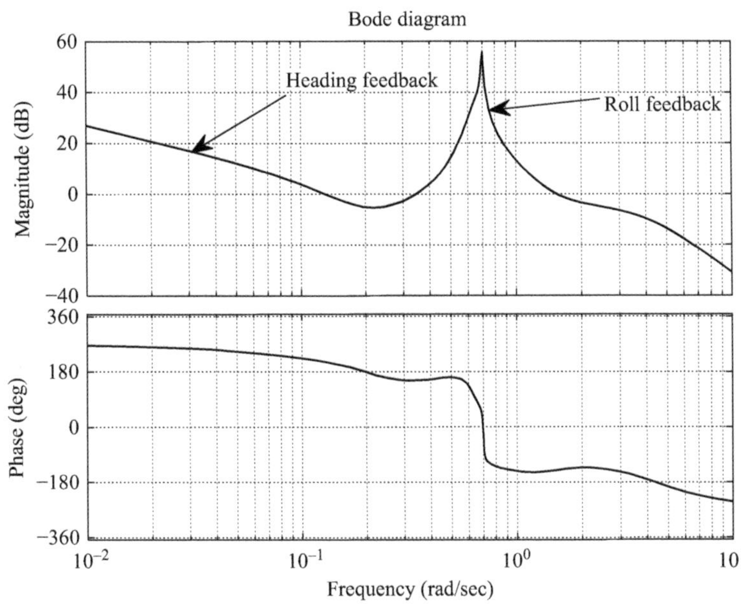

Figure 6.30 SITO return ratio

6.5.2 Case study: poor application of Nyquist-stable control

Another alternative to improve shaft rate regulation subject to the bandwidth limitations quantified in Section 6.1.4 is to have a steeper roll-off to the crossover frequency (i.e. roll-off steeper than −10 dB/oct). This provides greater negative feedback over the functional bandwidth. A *Nyquist-stable* control system is one that is stable in closed loop and has a stable loop transmission function that crosses the negative real axis of the Nyquist plot outside the unit circle. This describes a loop transmission function that over an interval of frequencies has a modulus slope steeper than −12 dB/oct. Nyquist-stable controllers have been implemented on aerospace plants with excellent results [10, 56]. The trade-off for this increased feedback is a susceptibility to oscillation in actuator saturation. Nonlinear dynamic compensation can be employed with a Nyquist-stable system to satisfy the conditions of absolute stability [10, 63]. This is discussed in the next chapter.

To test the efficacy of a Nyquist-stable design for the 5 MW wind turbine, a SISO collective blade pitch/high-speed shaft rate controller with this property is designed and tested in a turbulent wind environment. This controller has the advantage of the first tower mode control being abandoned (item 2 from Section 6.1.4 is no longer a factor), so the bandwidth can be increased to 1 rad/s. Figure 6.31 shows the Nichols plots without M-rings for the low-order high-speed shaft rate feedback only loop transmission and the Nyquist-stable controller.

The Nyquist-stable controller is extremely aggressive. In addition to the steep roll-off, the stability margins are smaller than those of the SITO rate regulation controller,

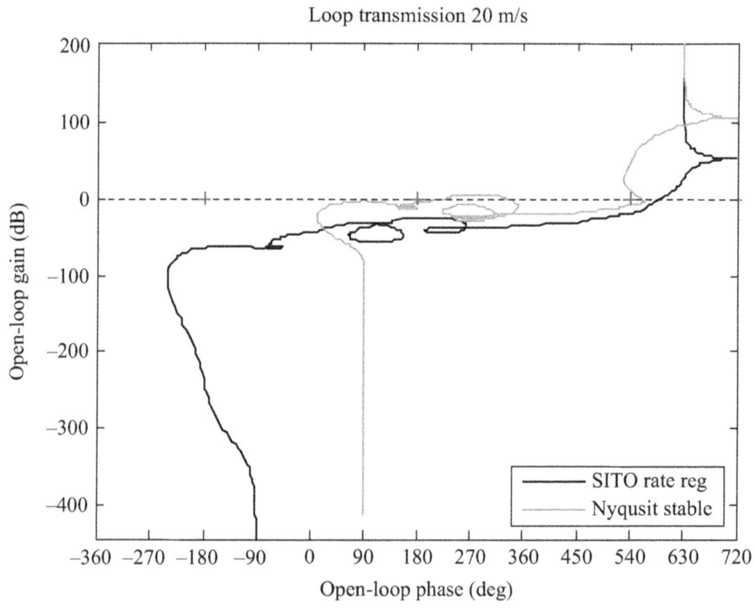

Figure 6.31 5 MW Nyquist-stable loop transmission Nichols chart

and the torsion mode is phase stabilized as opposed to gain stabilized. These features, in addition to the extra octave of bandwidth, give the Nyquist-stable controller approximately 40 dB more feedback at 0.1 rad/s than the SITO rate regulator.

The aggressive features of the Nyquist-stable control system result in poor robustness. This is a particularly bad characteristic for wind turbine applications, as the plant dynamics vary with wind speed. Figure 6.32 shows the Nyquist-stable loop transmission for four different wind speeds in Region 3. While retaining stability at wind speeds higher than the design speed (18 m/s), there are critical point encirclements at both the initial crossover and the phase-stabilized torsion mode. FAST simulations with this controller with turbulent wind frequently show violent oscillations within 1 min of operation.

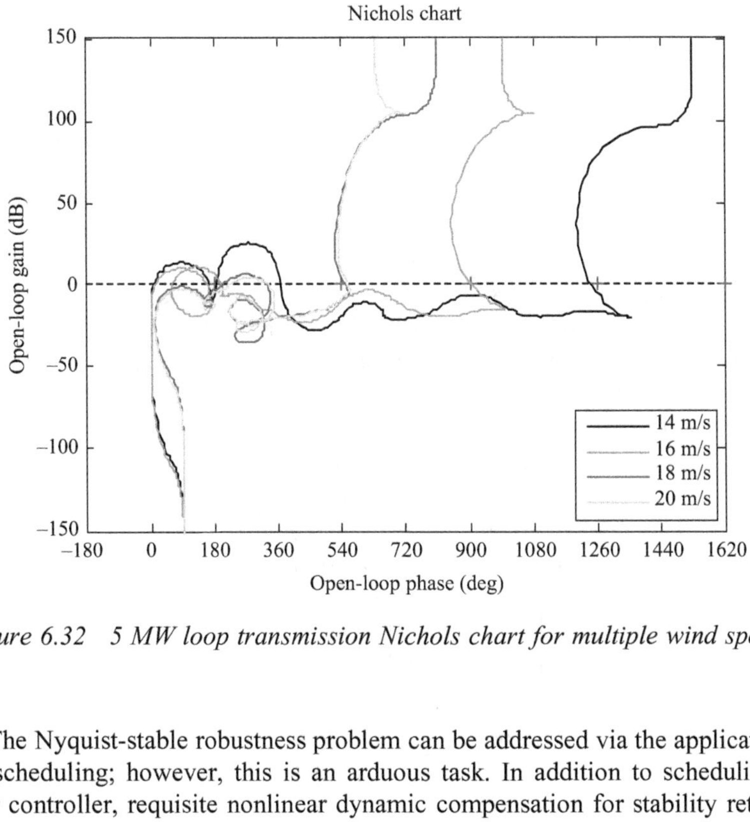

Figure 6.32 5 MW loop transmission Nichols chart for multiple wind speeds

The Nyquist-stable robustness problem can be addressed via the application of gain-scheduling; however, this is an arduous task. In addition to scheduling the linear controller, requisite nonlinear dynamic compensation for stability retention in actuator saturation would also have to be gain scheduled. The result would be a very complicated, perhaps fragile control system. In addition, the increase in performance at low frequency is offset by increased positive feedback near crossover. The advantages of Nyquist-stable control systems are realized when there is substantial feedback bandwidth, a characteristic that wind turbines do not have. Thus, this type of aggressive, nonlinear control (Nyquist-stable control with nonlinear dynamic compensation) is ill-suited for wind turbines.

6.6 Exercises

1. If the Bode minimum phase margin is 30 degrees, what is the best loop transmission modulus slope for a linear feedback system in the interval of frequencies between the functional bandwidth and the crossover frequency?
2. A control system is designed using the Bode optimal loop shape. The bandwidth is 200 rad/s. The functional bandwidth is 0.1 rad/s, with a subsequent roll-off slope of −8 dB/oct. What is the feedback over the functional bandwidth?
3. The plant is $P(s) = \frac{20{,}000}{s(s^2+s+100)}$. The loop transmission gain is flat to 0.1 rad/s. The 0 dB crossover is 60 rad/s. The high-frequency roll-off is −24 dB/oct. Design a Bode step at −8 dB for this controller. Report the feedback applied below 0.1 rad/s. Sketch the Bode and Nyquist plots of the loop transmission of the controller.
4. The phase shift for a minimum phase system at all frequencies is −200 degrees. What is the gain slope?
5. A feedback control system is designed with the Bode optimal cutoff and a Bode step. The gain and phase margins are 10 dB and 30 degrees, respectively. The return ratio phase at very high frequency (much higher than the end of the Bode step) is −270 degrees. The system has a time delay of 1 ms, causing 0.62 rad of nonminimum phase delay at the end of the Bode step. What is the crossover frequency of the control system?
6. The control system of the previous problem has 40 dB of feedback over its functional bandwidth. What is the functional bandwidth?
7. You are to design a tracking control system for a SISO 9th-order plant. The numerator and denominator polynomial coefficients in descending order are:

$$\text{Pnum} = 1.0e+016 \begin{bmatrix} 0.000000100000000 \\ 0.000100210000000 \\ 0.004220520000000 \\ 0.018922000000000 \\ 0.422850000000000 \\ 0.130000000000000 \\ 2.000000000000000 \end{bmatrix} \tag{6.20}$$

and

$$\text{Pden} = 1.0e+013 \begin{bmatrix} 0.000000000000100 \\ 0.000000000011110 \\ 0.000000401127310 \\ 0.000004542885340 \\ 0.004065517671600 \\ 0.005032607312000 \\ 0.608953907200000 \\ 0.094323200000000 \\ 4.608000000000000 \\ 0 \end{bmatrix} \tag{6.21}$$

The following are requirements. A functional bandwidth of 10 rad/s is required. The return ratio does not have to be smooth. Stability must be retained when the actuator is saturated. The controller must be linear. The transient response to step commands should be well damped. The effect of sensor noise should not result in more than 2% peak error in steady-state response to step command. The following are the details. There is a 0.5 ms time delay. The sensor noise power spectral density (PSD) is shown in Figure 6.33. You are to design a control system that satisfies the requirements subject to the details. Provide evidence of performance and robustness. Be quantitative.

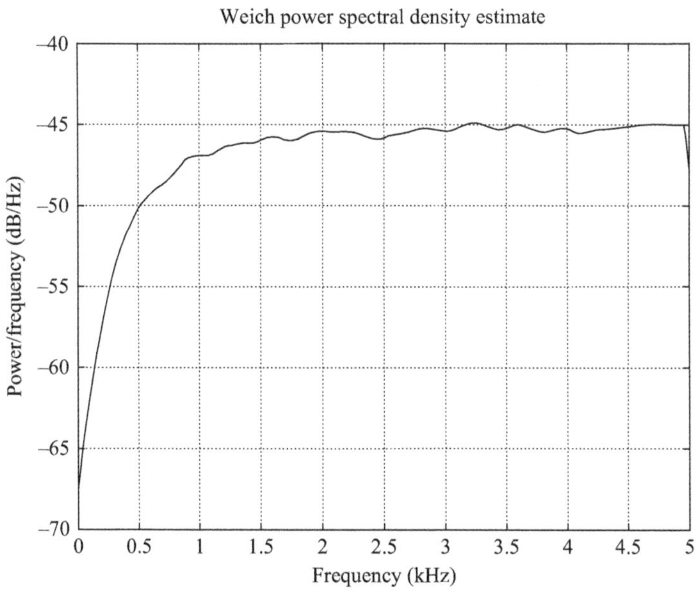

Figure 6.33 Sensor noise PSD

8. The LTI SISO plant transfer function is $P(s) = \frac{10{,}000}{(s+1)(s^2+2s+1600)}$. The feedback bandwidth is 1000 rad/s. Design a compensator $C(s)$ such that the functional bandwidth is 1 rad/s, the roll-off after 1 rad/s is −9 dB/oct, the gain margin is 9 dB, and the roll-off slope at very high frequency is −18 dB/oct.
9. For the plant of the previous problem, design a Nyquist-stable control system with the same bandwidth and relative stability. Compare the feedback over the functional bandwidth of both controllers.
10. Write a Simulink model for the controllers in Problems 2 and 3. The reference input is the unit step. Add a unit amplitude sinusoidal source to the input of the plant. This is the disturbance. Plot the responses to the step reference and sinusoidal disturbances at 0.5, 10, 100, 1000 and 2000 rad/s. Compare the results.

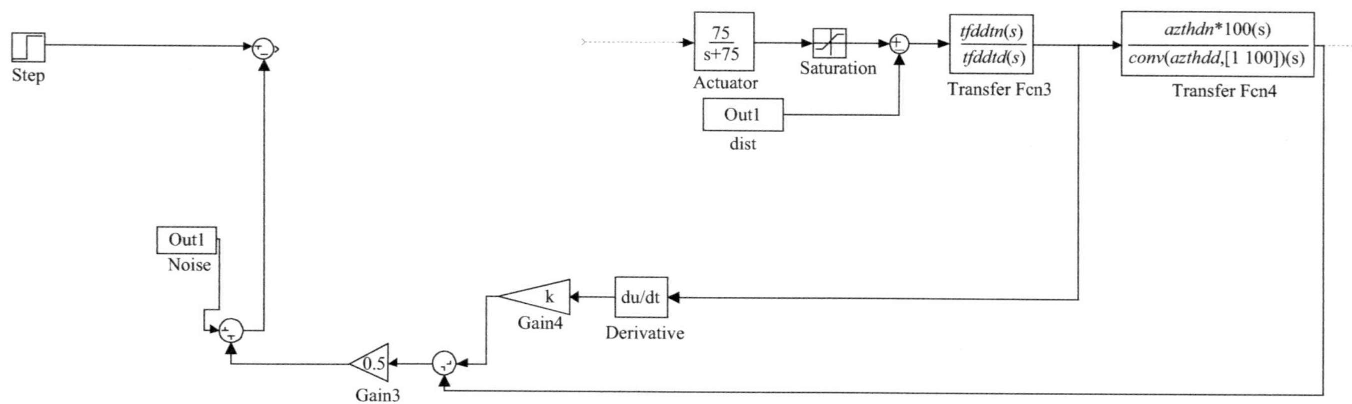

Figure 6.34   Missile autopilot design

11. The mode at 40 rad/s shifts to 80 rad/s. Quantify the effect on the controllers designed in Problems 2 and 3.
12. The LTI SISO plant transfer function is $P(s) = \frac{5 \times 10^7}{(s+1)(s^2+2s+1600)(s^2+s+4 \times 10^6)}$. Design a high-performance feedback control system with a functional bandwidth of 1 rad/s, and a 0 dB crossover frequency of 500 rad/s. Use phase stabilization.
13. Design an autopilot for a missile whose dynamics are linearized at a particular operating point. The block diagram is shown in Figure 6.34 [77]. The missile should track a 20 g command as fast as is practical, with good transient response characteristics. Atmospheric disturbances should be well rejected. There should not be excessive tail fin saturation. Do not pole-zero cancel the short period mode. The influence of sensor noise should not be excessive. Provide a complete Simulink model of your design. Provide the loop transmission function plot for your design. Provide a time response of the closed-loop system (accelerometer output) to the 20 g command with the disturbance and noise inputs applied. The noise and disturbance models are shown in Figure 6.35. The saturation limits are ± 20 degrees. The sensor conversion constant is $k = -0.0012$. Plant polynomials are $tf\,ddtn(s) = -469.6s + -563.52$, $tf\,ddtd(s) = s^2 + 1.27s + 72.25$, $azthdn(s) = 0.0021s^2 - 2.4273$ and $azthdd(s) = s + 1.2$.

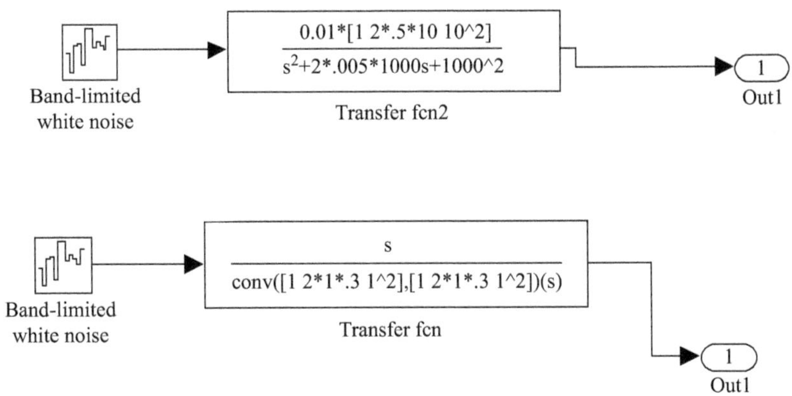

Figure 6.35 Missile autopilot noise (top) and disturbance (bottom) models. The noise power is 0.001 with a sample time of 0.01 s. The disturbance power is 10 with a sample time of 0.01 s.

14. Write a Simulink model to determine the effect of saturation in the actuator 1 in the TISO system described in the example in Section 6.4.
15. The plant is $P(s) \in \mathbf{F}^{2\times 1}(s)$, $P(s) = \left[\frac{100}{s(s+1)} \; \frac{1000s^2}{s^3(s^2+s+40,000)}\right]^T$. Design a regulator for this system. Plot the loop transmission function. Verify sufficient frequency separation.

Chapter 7
Feedback design – nonlinear

> The controller has more than 100 dB of feedback...but please don't touch the plant!
>
> – *Anonymous*

Frequency-domain techniques for high performance control are introduced in the previous chapter for single-input, single-output (SISO), single-input, two-output (SITO) and TISO architectures. Taking into consideration the bandwidth limitations presented by all practical control design problems, these techniques provide more feedback (thus performance) than commonly used controllers, like proportional-integral-derivative (PID). The principle trade-off is increased compensator order, which is very inexpensive and well worth the controller performance improvement.

It is assumed that the designs discussed in Chapter 6 are applied to *linear* systems. Although the control designer may be confident in the accuracy of a linearized model of the system to be controlled, all actuators eventually saturate. The more aggressive control strategies presented in Chapter 6 drive the actuators harder than less aggressive controllers, thus increasing the probability of saturation. This is particularly true of the Nyquist-stable controller that lacks absolute stability (AS) in the sector $[0 \quad 1]$ and is all but guaranteed to go unstable in saturation. These types of control systems should never be deployed without nonlinear compensation described in this chapter.

In this chapter, nonlinear methods are discussed to improve the performance of linear controllers with nonlinearities in the loop. It is the incorporation of these techniques with those discussed in Chapter 6 that deliver good disturbance rejection in the small-signal condition, and stability retention in the large-signal condition.

7.1 Anti-windup

Consider a proportional-integral (PI) compensator ($C(s)$) with a saturation in the forward path (block diagram shown in Figure 7.1). The saturation limits are those of the actuator. When the actuator saturates, $e(t)$ builds, and the signal in the integral branch of the compensator increases until the error changes sign. This is called

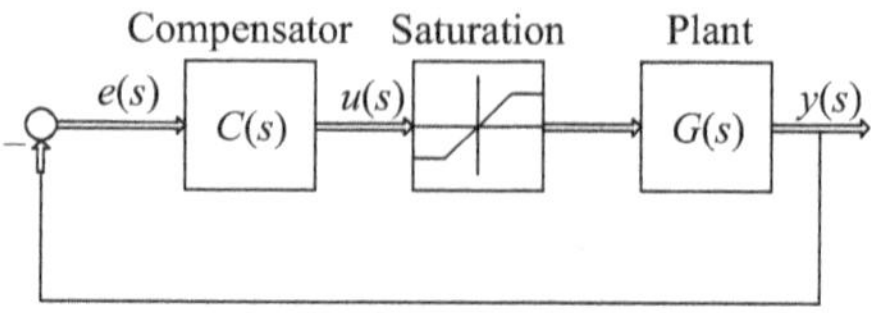

Figure 7.1 PI compensator with actuator saturation

integrator windup and is the reason why controllers with integral action recover sluggishly from saturation.

The *anti-windup* control input from the integral branch is $u_i = K_I \int_0^t e(t)dt + K_{aw} \int_0^t e_s(t)dt$, where K_I is the integral gain and K_{aw} is the anti-windup gain. Signal e_s is the difference between the input and the output of the saturation block in the forward path. When the input is within the range of the actuator, $e_s = 0$, the control is the nominal PI controller. Outside the limit, the sign of e_s is opposite the saturated direction, and the integral action is reduced. The recovery time is thus reduced.

7.2 Nonlinear dynamic compensation

Nyquist-stable control systems, discussed in Chapter 6, are high-performance systems but are susceptible to oscillation in saturation. To be feasible in a practical sense, the Nyquist-stable controller must be augmented with nonlinear compensation.

Consider a feedback loop with the actuator mathematically modeled by nonlinear function $A(u)$. It is not possible to use linear control theory like the Nyquist Criterion to determine the stability of this feedback system. Figure 7.2 shows a modified system with an additional feedback path with a nonlinear link. This is a *nonlinear dynamic compensator* (NDC). When $A(u) = u$, the signal through the new feedback path is zero, and the loop transmission is $T(s) = C(s)P(s)$. When $A(u) \neq u$, the signal through the new feedback path is nonzero and filtered through N_1.

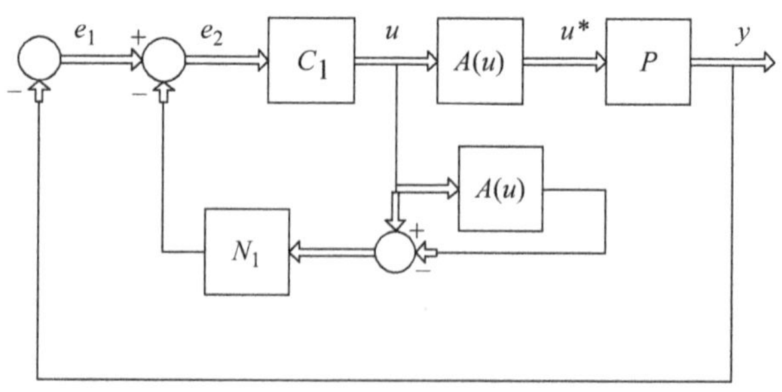

Figure 7.2 The nonlinear dynamic compensator

The equivalent linear system in feedback with the saturation is as follows:

$$T_{eq} = \frac{P(s)C(s) - N(s)C(s)}{1 + N(s)C(s)} \tag{7.1}$$

This system is in feedback connection to the nonlinearity, as depicted in Figure 7.3. The system $N(s)$ is be designed such that the *equivalent system* $T_{eq}(s)$ satisfies the conditions of absolute stability.

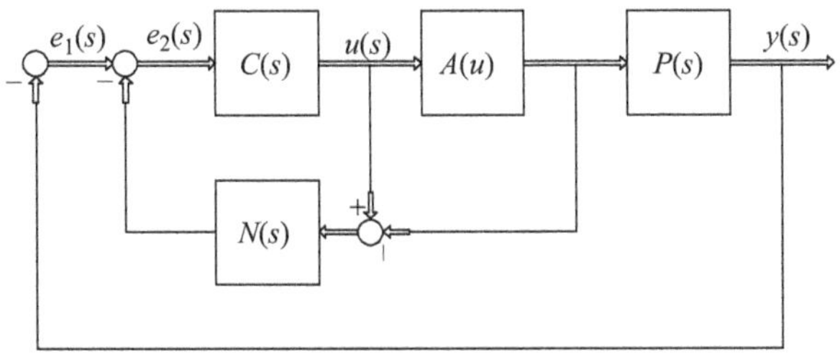

Figure 7.3 Equivalent block diagram of feedback loop with NDC

When $A(\cdot)$ is the saturation nonlinearity, a very common case, $T(s)$ is the small signal loop transmission (in the linear range of the actuator). If the deadzone interval in the feedback path is equivalent to the linear interval of the saturation in the forward path, there is an equivalent feedback loop of Figure 7.4. The NDC and the Nyquist-stable control system is a very effective combination. In the small-signal condition, the large feedback of the Nyquist-stable controller delivers excellent disturbance rejection. The NDC retains stability when the actuator saturates, a characteristic the Nyquist-stable system alone does not possess.

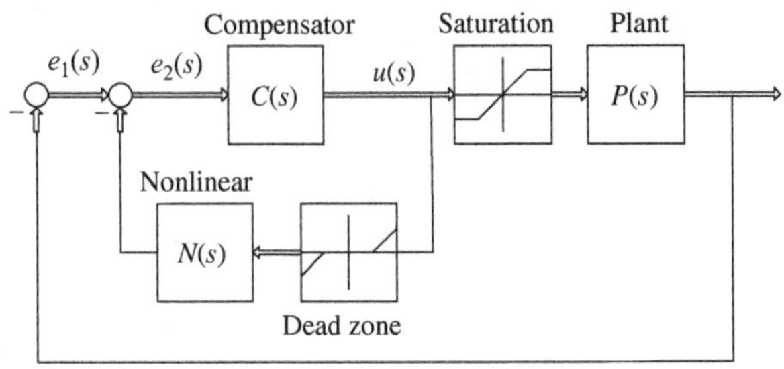

Figure 7.4 Block diagram of a system with actuator saturation and NDC

7.2.1 Case study: nonlinear dynamic compensator design for a vibration suppression system

The plant is a *PUS* – *RR* parallel robot shown in Figures 7.5 and 7.6. One kinematic path consists of an active prismatic actuator, universal joint, and sphere joint (*PUS*), whereas the other is an active revolute joint in series with a pin joint (*RR*).

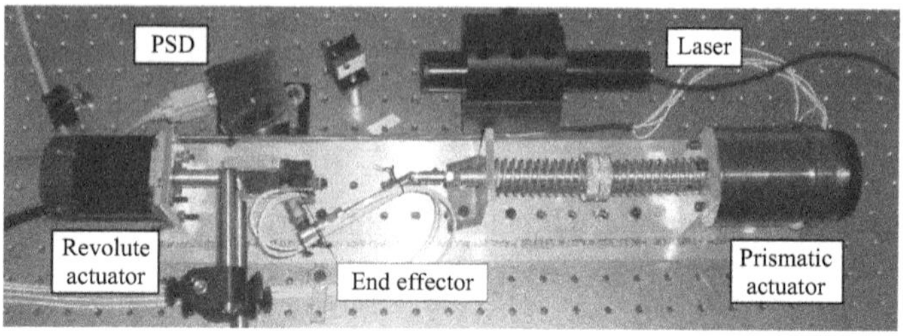

Figure 7.5 $\underline{P}US - \underline{R}R$ parallel robot

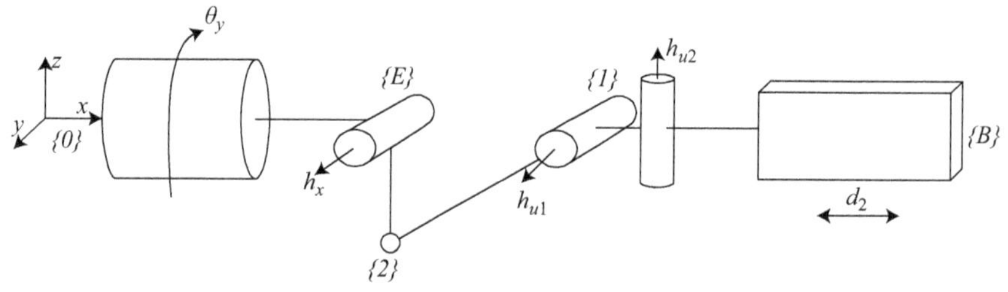

Figure 7.6 $\underline{P}US - \underline{R}R$ parallel robot schematic

Actuation
The prismatic actuator is powered by a linear voice coil manufactured by BEI Kimco Magnetics Division, model number LA28-43-000A. It is capable of approximately 300 N of force and 3 cm of displacement. The coil is mechanically in series with helical springs that provide a restoring force to keep the end-effector centered when unpowered. The coil is driven by an AE Techron LV608 Linear Power Amplifier operating in a voltage-controlled voltage source mode with a gain of 30 V/V. The revolute actuator is a standard NEMA 23 sized electric motor by Techic driven by an SSt-Eclipse digital controller running in torque control mode. This DOF is not used for the control experiments described in this paper.

Sensing
Attached to the end-effector are two microelectromechanical system (MEMS) gyros used to measure the angular rate about the two DOFs. These Analog Devices ADXRS610 300° single-axis gyroscope sensors have an operational bandwidth of 360 Hz. The quiescent noise power spectrum of the gyro, with a peak at 14 kHz, can be seen in Figure 7.7. A time delay of approximately 0.75 ms $\approx 1/(1.4$ kHz) is observed. Such nonminimum phase delay limits the attainable bandwidth and thus is not used for closed-loop experimentation.

A Helium Neon (HeNe) laser is used in conjunction with an ON-TRAK Photonics OT-324 Position Sensitive Detector (PSD) to measure the end-effector position. The stationary HeNe laser emits a 632.8 nm beam that is directed by an adjustable steering mirror toward the end-effector. A mirror attached to the end-effector deflects the beam onto the bare duolateral photodiode sensor of the PSD. The PSD sensing amplifier reports the position of the laser beam on the

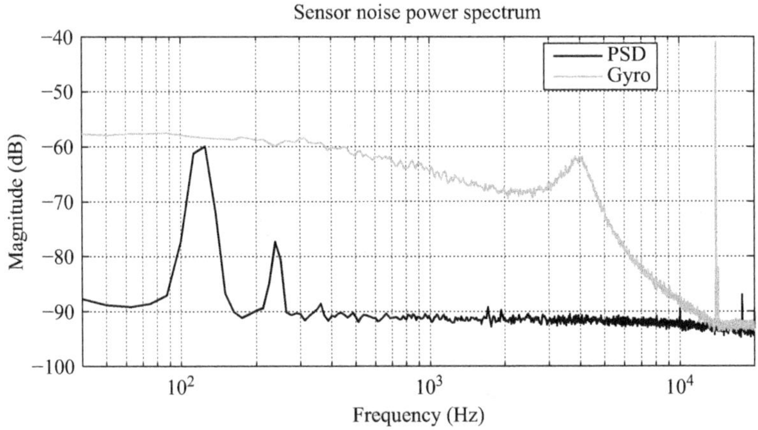

Figure 7.7 Quiescent sensor noise power spectrum densities

two-dimensional sensor as two analog voltages within ±10 V. Figure 7.7 shows the quiescent noise power spectrum of the position sensor along one axis. There are peaks in the position sensors spectra at 120, 240, and 360 Hz. These harmonics are likely caused by a 120 Hz voltage ripple on the high-voltage DC line powering the laser that results from rectifying the 60 Hz AC power. The higher operational bandwidth of nearly 15 kHz and lower noise characteristics make the PSD better suited than the gyro for the control experiments discussed in this paper.

System identification

The system is driven by the prismatic actuator with band-limited Gaussian noise for the purpose of system identification. The drive and PSD signals are digitized and recorded using an Agilent VXI system. The data capture and processing is performed using the SignalCalc 6.2 software suite from DataPhysics. Stimuli of varying amplitudes and bandwidth are applied and the corresponding coherence and plant transfer functions are calculated. A nominal pole-zero-gain (PZK) model of the plant is developed by the manual placement of poles and zeros to sufficiently match those transfer functions with the highest coherence. This 15th-order nominal model $P(s)$ is plotted along with the experimentally acquired data in Figure 7.8. The gain zeros and poles of $P(s)$ are $K = 4.4 \times 10^{20}$, $s_z = (-5027, -2.482 \pm 496.4i, -904.8 \pm 2877i, -108.1 \pm 5402i)$ and $s_p = (-25.13, -24.82 \pm 495.8i, -301.6 \pm 402.1i, -1319 \pm 2285i, -184.7 \pm 2632i, -257.6 \pm 5146i, -457.4 \pm 6518i, -326.7 \pm 8162i)$.

The plant transfer function is low-pass as expected. The plant modulus is relatively flat with a value around 48 dB from 0 Hz to around 4 Hz, at which point it begins to roll off at approximately 6 dB/oct. At 80 Hz, the slope transitions to nearly 18 dB/oct, as there are two conjugate pole pairs and one very lightly damped conjugate zero pair around this frequency. There are interlaced pole/zero pairs around 440 and 840 Hz. There are consecutive pole pairs at 1 and 1.3 kHz, whose

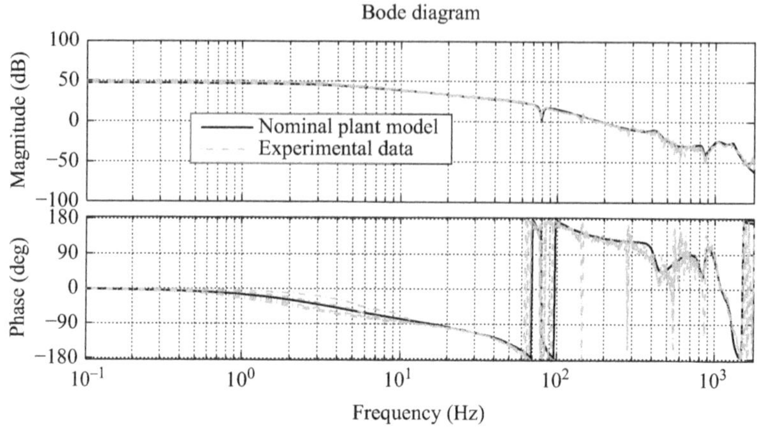

Figure 7.8 Actual and modeled plant frequency responses

additional total phase contribution of −360 degrees places a hard upper limit on control bandwidth and thus the amount of available feedback. The phase contributions of the well-damped poles and lightly damped poles around 80 Hz likewise limit control bandwidth as described.

Absolutely stable fixed gain controller design

An absolutely stable fixed gain (ASFG) controller is designed to serve as a baseline for comparison to the nonlinear controller described later. The design goal is to maximize vibration suppression at frequencies below 10 Hz while remaining absolutely stable. This is accomplished by carefully shaping the compensator $C_{ASFG}(s)$ to obtain the return ratio $T_{ASFG}(s) = C_{ASFG}(s) \cdot P(s)$ seen in Figure 7.9. Similar to the development of the nominal plant model, the PZK model for C_{ASFG} is designed by manually placing poles and zeros (Figure 7.10).

A zero at 4 Hz and a conjugate pair of zeros at 80 Hz compensate for the corners found in the original plant. A pair of conjugate poles placed at 11 Hz define the corner of the functional bandwidth around 10 Hz. A pair of zeros at 100 Hz together with pairs of poles at 160 and 220 Hz define the Bode step. A lead filter with a zero at 60 Hz and pole at 100 Hz, together with the Bode step, provide sufficient phase advance around the crossover point. Thus, a gain margin of at least 10 dB and a phase margin of nearly 50 degrees is maintained. This may seem excessive, but it allows the demonstration of absolute stability (AS) shown later. The gain, zeros and poles of $C_{ASFG}(s)$ are $K = 196$, $s_z = (-25.13, -377, -301.6 \pm 402.1i, -502.7 \pm 377i)$ and $s_p = (-628.3, -41.47 \pm 55.29i, -502.7 \pm 870.6i, -414.7 \pm 1319i)$.

A major factor limiting bandwidth is the conjugate zeros near the imaginary axis and heavily damped poles around 80 Hz. The phase lag of the poles is more pronounced than the phase lead of the zeros at frequencies below 80 Hz. Thus, the phase sharply drops to nearly −180 degrees around 80 Hz. There is also uncertainty in the exact frequency and Q factor of this mode, as it varied between different

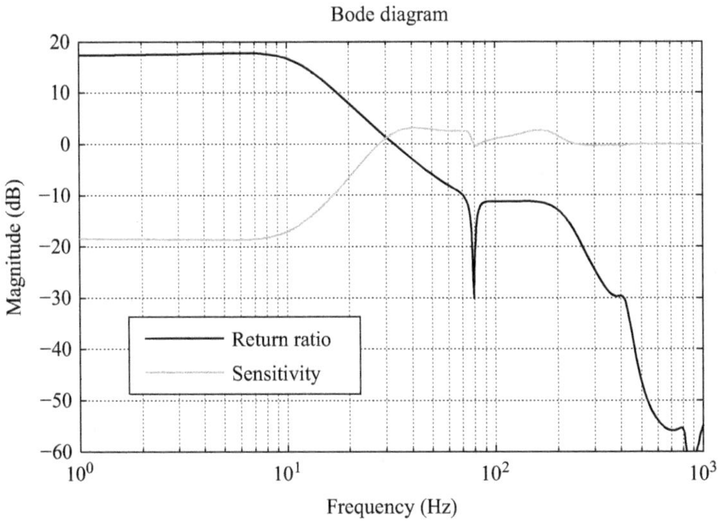

Figure 7.9 ASFG controller design

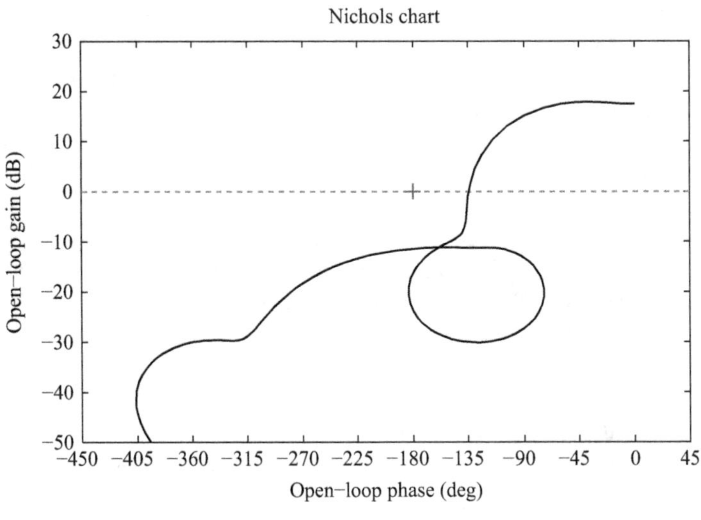

Figure 7.10 ASFG controller design

system identification experiments. Thus, the mode is gain stabilized in order to retain adequate phase margins. The return ratio crosses 0 dB around 33 Hz and the resulting 7th-order controller provides nearly 18 dB of negative feedback below 10 Hz.

Return ratios T_{ASFG} and sensitivities $S_{ASFG} = 1/(1+T_{ASFG})$ are also calculated using the experimentally found plant response data to verify that the final compensator design is acceptable when the plant differs from the nominal model. In all cases, stability is retained and the modulus of positive feedback (where $|S_{ASFG}| > 0$ dB) is less than 5 dB. Absolute stability of the ASFG controller is shown in later.

Nyquist-stable controller design

A high-performance Nyquist-stable (NS) controller $C_{NS}(s)$ is also designed with the goal of maximizing disturbance rejection at frequencies below 10 Hz. By definition, the return ratio of a Nyquist-stable system has a modulus greater than unity and phase less than 180 degrees over some interval of frequencies. Furthermore, since the plant is open loop stable, the Nyquist plot of the return ratio $T_{NS}(s) = C_{NS} \cdot P(s)$ must not encircle the -1 critical point. As opposed to the ASFG, however, the NS controller by itself is not required to be AS.

C_{NS} is designed by carefully shaping the return ratio by direct placement of poles and zeros. A zero at 4 Hz compensates for the bend in the original plant. Conjugate pole pairs were placed at 6.75 and 10 Hz to fix the functional bandwidth at 10 Hz. The pole pair at 10 Hz and a zero pair at 15 Hz were lightly damped in order to provide the sharp transitions seen in the slope of the return ratio at 10 and 15 Hz. It is this steep slope and accompanying phase lag that allows us to attain nearly 20 dB more negative feedback (where $|S_{NS}| < 0$ dB) than with the ASFG controller. A pair of zeros at 60 Hz and a pair of poles at 200 Hz flatten the response to produce a Bode step. A two-octave-wide lead filter centered at 100 Hz further flattens the response and adds phase lead. The loop is shaped such that the positive feedback never exceeds 6 dB. The gain, zeros and poles of $C_{NS}(s)$ are $K = 2119$, $s_z = (-25.13, -314.2, -14.14 \pm 93.18i, -188.5 \pm 326.5i)$ and $s_p = (-1257, -33.93 \pm 25.45i, -12.57 \pm 61.56i, -1257 \pm 2177i)$.

The resulting 7th-order compensator provides nearly 38 dB of disturbance rejection below 10 Hz as is evident in the Bode plot of the loop transmission shown in Figure 7.11. This is an order of magnitude greater than the ASFG controller. The 0 dB crossover frequency is also slightly higher at 44.5 Hz. As can be seen in Figure 7.12, the return ratio satisfies the Nyquist stability criterion and the system without saturation is stable in closed loop. However, no physically realizable system behaves linearly under all conditions. The saturation of large signals is inevitable considering the finite nature of real-world systems.

Effects of saturation

The plant possesses several saturation mechanisms: the controller analog output voltage is limited to ± 10 V, the maximum power output of the Techron amplifier is only a few hundred watts, the stroke of the prismatic actuator is physically limited, and the laser beam can only be deflected a few degrees before it misses the target PSD sensor. To investigate the effects of saturation on the closed-loop system without driving the hardware to its physical operating limits, an additional saturation

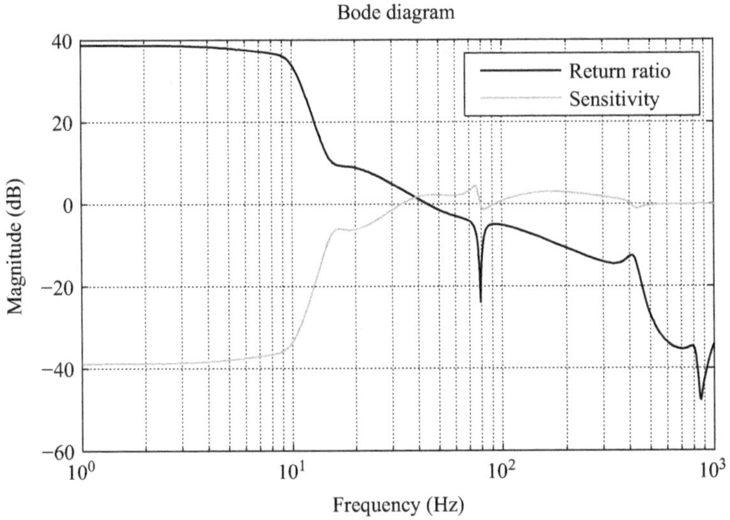

Figure 7.11 NS controller design

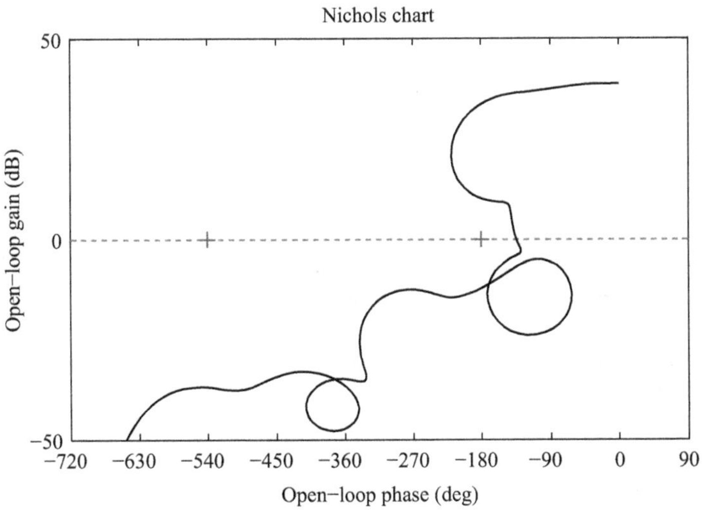

Figure 7.12 NS controller design

with an adjustable threshold is added to the controller output. This system is shown in Figure 7.13. This hard saturation is described by the equation

$$S(x) = \begin{cases} t_{sat} \cdot sign(x) & : |x| > t_{sat} \\ x & : otherwise \end{cases} \tag{7.2}$$

where $S(x)$ is the saturation link output, x is its input, and t_{sat} is the adjustable saturation threshold. The values of t_{sat} used during closed-loop experimentation are sufficiently low so that this link will saturate before any of the other aforementioned sources of saturation.

Consider a sinusoid of amplitude E fed into a saturation block. For a small-amplitude sinusoid, $E < t_{sat}$, the saturation link is equivalent to a unity gain. For a large-amplitude sinusoid, $E > t_{sat}$, the output is a waveform of reduced amplitude but of the same frequency and phase of the input. Thus, a saturation link can effectively reduce the loop gain for large signals.

Consider the Nichols plots for the ASFG and NS controllers in Figures 7.10 and 7.12. A broadband reduction of the return ratio moduli would result in the plots shifting down along the amplitude axis. Since $|T_{NS}| > 0$ dB at some frequencies where $arg(T_{NS}) < -180$ degrees, reducing the gain will eventually result in an encirclement of the critical point and the Nyquist Criterion would no longer be satisfied. This is not the case for the ASFG controller.

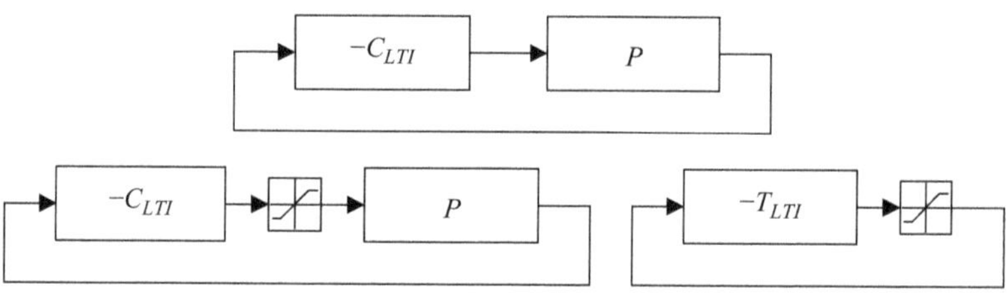

Figure 7.13 Standard feedback configurations and equivalent system used for absolute stability analysis

Absolute stability and the Popov Criterion

A linear, time invariant (LTI) system in feedback with a nonlinearity $n(e)$ is said to be AS if it is asymptotically globally stable for any $n(e)$ satisfying the sector condition $0 < e \cdot n(e) < e^2$. The saturation $S(x)$ in (7.2) clearly satisfies the sector condition, so absolute stability is certainly applicable to any system equivalent to that seen in Figure 7.13.

The Popov Criterion is a *sufficient* condition for absolute stability. The criterion requires that the return ratio $T(s)$ have no poles in the closed right half plane, and that there exists some $q > 0$ such that $Re\{(1+jq\omega) \cdot T(j\omega)\} > -1$ for all ω. If these conditions are met, then the closed-loop system is AS. This inequality can equivalently be written as $\omega ImT < q^{-1}(ReT+1)$, which provides for a simple graphical interpretation. The Popov line passing through the critical point with slope q^{-1} satisfies the *equality* $\omega ImP_L = q^{-1}(ReP_L+1)$. Therefore, the inequality holds if the plot of $ReT + j\omega ImT$ on the modified Nyquist plane lies entirely to the right of some Popov line with positive slope.

All of the poles of T_{ASFG} have negative real parts and the left plot of Figure 7.14 clearly shows the modified Nyquist plot for T_{ASFG} lying to the right of a Popov line with slope $q^{-1} = 375$. Thus, the Popov criterion is satisfied and the ASFG control system truly is absolutely stable. T_{NS} is also open loop stable, but by

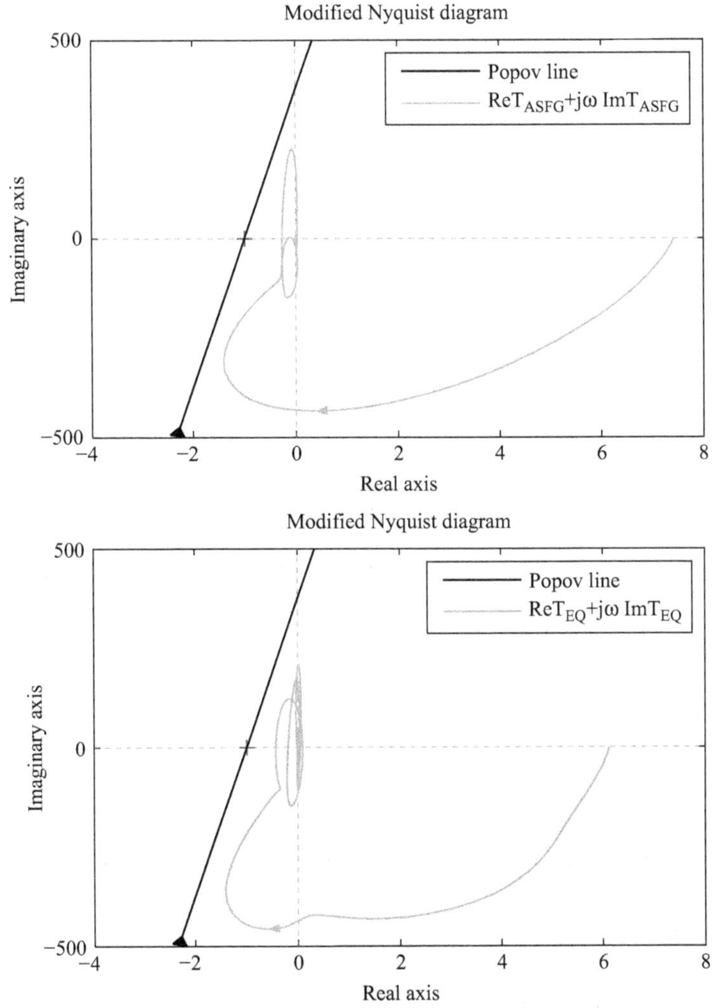

Figure 7.14 Modified Nyquist plots for absolute stability analysis

the definition of Nyquist stable there exists an ω_0 such that $arg(T_{NS}(j\omega_0)) = -180$ degrees and $|T_{NS}(j\omega_0)| \geq 1$. Therefore, $Re[(1+qj\omega_0)T_{NS}(j\omega_0)] = -|T_{NS}(j\omega_0)| \leq -1$ and the NS control system cannot satisfy the Popov criterion.

Nonlinear dynamic compensator design

An adaptation of the control scheme in Figure 7.13 is shown in Figure 7.15. The additional feedback path is the nonlinear dynamic compensator (NDC). For small signals, the output of the summing junction in the NDC path is zero. Thus, the effective return ratio for small signals is simply T_{NS}. For large signals, however, this difference is nonzero and the NDC feedback path is active. Therefore, the NDC is able to reshape the effective return ratio of the controller for large signals.

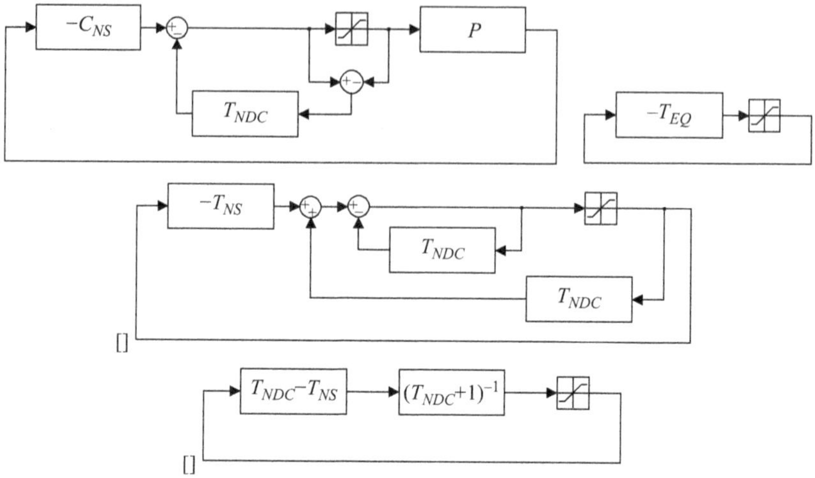

Figure 7.15 Equivalent systems for the NS with NDC controller

Figures 7.15 illustrates a series of loop transformations used to find an equivalent system necessary for AS analysis. The return ratio of the equivalent system T_{EQ} is found to be

$$T_{EQ} = \frac{T_{NS} - T_{NDC}}{1 + T_{NDC}} \tag{7.3}$$

where T_{NDC} is the return ratio of the local NDC feedback loop and T_{NS} is the return ratio of the NS controller. The purpose of the NDC is to obtain an AS equivalent return ratio T_{EQ}. Thus, T_{ASFG} can serve as a target for T_{EQ}. For a given desired equivalent return ratio $T_{EQ}^D = T_{ASFG}$ and T_{NS}, a target NDC design T_{NDC}^D can be found by

$$T_{NDC}^D = \frac{T_{NS} - T_{EQ}^D}{1 + T_{EQ}^D} = \frac{T_{NS} - T_{ASFG}}{1 + T_{ASFG}} \tag{7.4}$$

The order of the resulting T_{NDC}^D is quite high, and its implementation could pose serious challenges. Therefore, a lower-order NDC is sought. A reduced-order NDC is thus developed by the manual placement of poles and zeros. An iterative approach is indicated here as the resulting equivalent return ratio T_{EQ} will differ from T_{EQ}^D, and AS will have to be verified. The resulting 9th-order T_{NDC} and 40th-order T_{NDC}^D are shown in the left plot of Figure 7.16.

The gain, zeros and poles of T_{NDC} are $K = 1.66 \times 10^{12}$, $s_z = (-1777, -46.37 \pm 61.83i, -37.7 \pm 119.9i)$ and $s_p = (-222.1, -33.93 \pm 25.45i, -15.71 \pm 60.84i, -942.5 \pm 1257i, -377 \pm 2485i)$. Well-damped pole pairs at 6.75 and 250 Hz in combination with lightly damped poles at 10 and 400 Hz capture the sharp bends in T_{NDC}^D seen around these frequencies. Likewise, well-damped zeros at 12.3 Hz and lightly damped zeros at 20 Hz produce the remaining corner. A pole around 35 Hz together with a zero around 282 Hz matches the modulus slope between these frequencies.

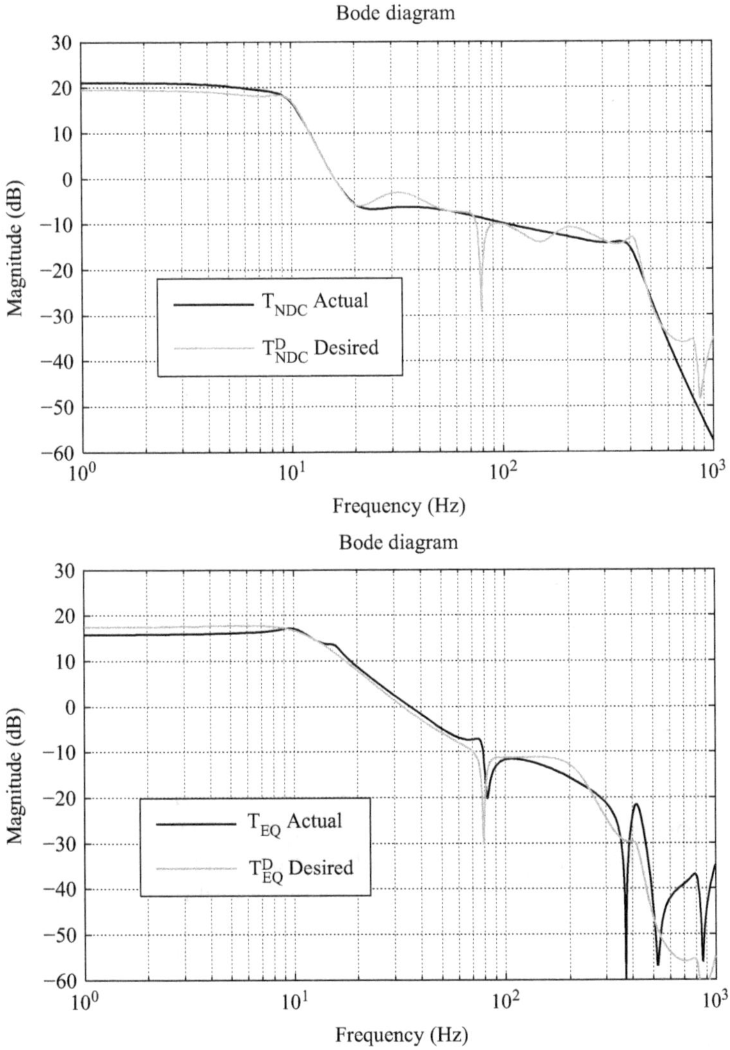

Figure 7.16 NDC design

The equivalent return ratio T_{EQ} of the reduced-order T_{NDC} and T_{NS} is calculated using (7.3), and plotted together with T^D_{EQ} in the right plot of Figure 7.16. The modified Nyquist plot of T_{EQ} is shown in the right plot of Figure 7.14 and lies entirely to the right of the Popov line with slope $q^{-1} = 375$. T_{EQ} is Hurwitz; therefore, the combined NS with NDC controller is AS.

Controller implementation

All control design work is performed using MATLAB and the Control Systems Toolbox. The continuous time PZK models of the final control designs are discretized using Tustin's method. This bilinear transformation uses the approximation

$$z = e^{sT} = \frac{e^{sT/2}}{e^{-sT}} \approx \frac{1 + sT/2}{1 - sT/2} \tag{7.5}$$

to map poles and zeros from the s-plane into the z-plane, where T is the sampling time. The poles and zeros were grouped to form a cascade of second-order systems. This approach of PZK design and CSOS implementation lessens the effects of finite precision arithmetic and round-off errors.

Custom software developed with National Instruments LabWindows/CVI executes on a National Instruments PXI 8106 Real Time Controller. The program implements the 7th-order C_{ASFG}, 7th-order C_{NS} and 9th-order T_{NDC} at a loop rate of 10 kHz. It should be noted that digitizing, processing and outputting of signals must take no longer than 1/10 kHz = 0.1 ms. This worst-case time delay is included in the controller models during the design phase. However, a 0.1 ms time delay only contributes 3.6 degrees of nonminimum phase at 100 Hz and 36 degrees at 1 kHz. Thus, it is not a major limiting factor in maximizing the crossover frequency. Companion software running on a Windows XP machine allows for loops to be open and closed, disturbances enabled or disabled and saturation thresholds and proportional gains to be adjusted in real time.

Disturbance environment

Although it would be desirable to mount the mechanism on a shake table, the facility does not currently have this hardware. The robot is rigidly mounted to an air suspended optics table, so the prismatic actuator itself is used for disturbance injection. A disturbance signal is simply added to the output of the controller before it is fed to the Techron amplifier input. A plot of the disturbance generating signal and its power spectral density are shown in Figure 7.17. The signal is generated by passing white noise through a 10 Hz first-order, low-pass filter and summing it with a 5 Hz sinusoidal tone.

Closed-loop performance

The performance of both the ASFG and the NS with NDC controllers is verified in closed-loop tests. The disturbance source described above is used in all cases, but t_{sat} is varied. A threshold of 150 mV is sufficiently high to ensure that control effort saturation seldom occurs. In this small-signal regime, the NDC is inactive and the effective return ratio of the NS with NDC controller is T_{NS}. This is evidenced by nearly 40 dB of disturbance rejection below 10 Hz as seen in the power spectral density in the left plot of Figure 7.18. A saturation threshold of 50 mV is sufficiently low to saturate the control effort some of the time. In this large signal scenario, the NDC link is active and reduces the effective return ratio to a less aggressive loop shape as shown in Figure 7.9. Comparing the performance of the ASFG controller between the small- and large-signal cases, we see that its effectiveness was also reduced in the saturated case as seen in the right plot of Figure 7.18.

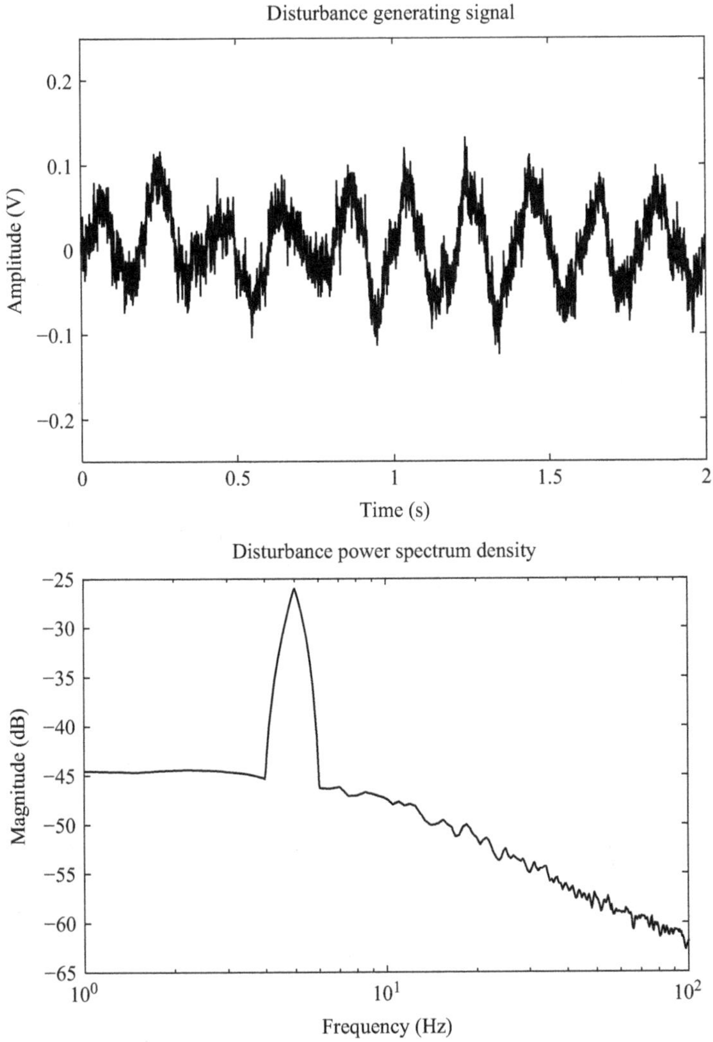

Figure 7.17 Disturbance generating signal used in closed-loop tests

The first half of Figure 7.19 demonstrates the unstable nature of the NS controller *without* the NDC. With $t_{sat} = 100$ mV, the NS controller without NDC is able to operate stably for some time until the control effort saturates. After this initial saturation, the control output oscillates wildly, ensuring further saturation. Stability can be regained by closing the NDC feedback path as shown in Figure 7.19 around the 1 s mark.

In practical applications, a Nyquist-stable control system should never be implemented with linear compensation only. This system is all but guaranteed to oscillate in saturation. As such, it should always be designed with nonlinear dynamic compensation to satisfy the conditions of absolute stability.

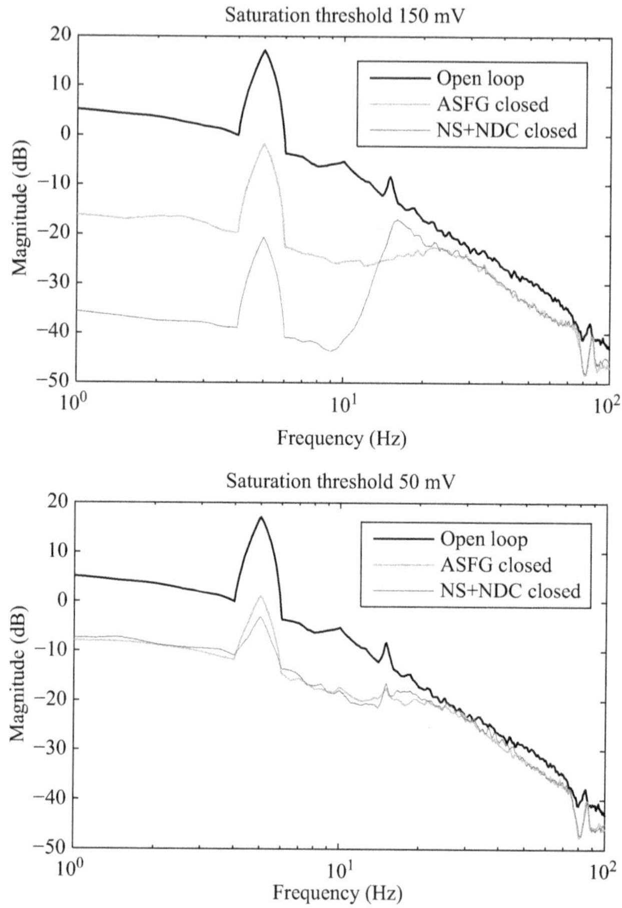

Figure 7.18 Open- versus closed-loop power spectrum densities

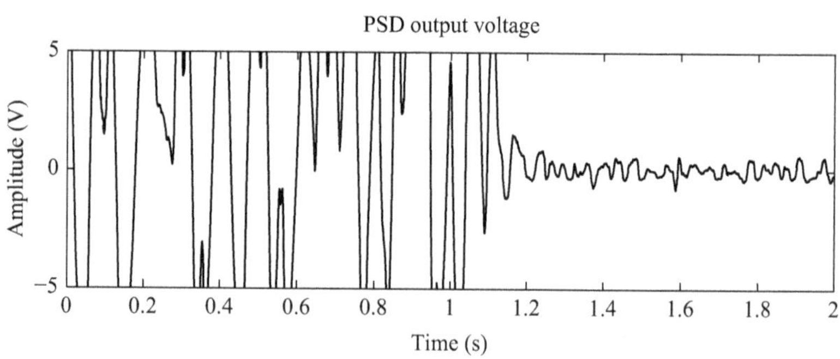

Figure 7.19 PSD sensor output showing NS controller regaining stability after NDC loop is enabled at around 1 s

7.3 Multipurpose nonlinear dynamic compensation

The previous section covers the use of nonlinear dynamic compensation to provide absolute stability in the presence of a sector nonlinearity in the feedback system. Multiple NDCs are now considered. The plant/actuator combination may have several unrelated nonlinearities that can destabilize the system or substantially reduce performance of systems where aggressive loop shaping is applied.

Consider the SITO feedback system shown in Figure 7.20. $A(\cdot)$ is a function operator on the actuator command. The NDC has multiple feedback paths through nonlinear functions $B(\cdot)$ and $C(\cdot)$ that satisfy the following conditions:

$$u_0 = \{u \in \Re : u^* = A(u) = u = B(u) = C(u)\} \tag{7.6}$$

$$u_1 = \{u \in \Re : u^* = A(u) = B(u) \neq u = C(u)\} \tag{7.7}$$

$$u_2 = \{u \in \Re : u^* = A(u) = C(u) \neq u = B(u)\} \tag{7.8}$$

The set of inputs is the union of these three mutually exclusive sets. The 'separation' of nonlinear conditions (i.e. inputs in u_1 or u_2 results in NDC feedback around *or* C_1 or C_2.) If $u \in u_0$ there is no signal through either of the NDCs. If $u \in u_1$,

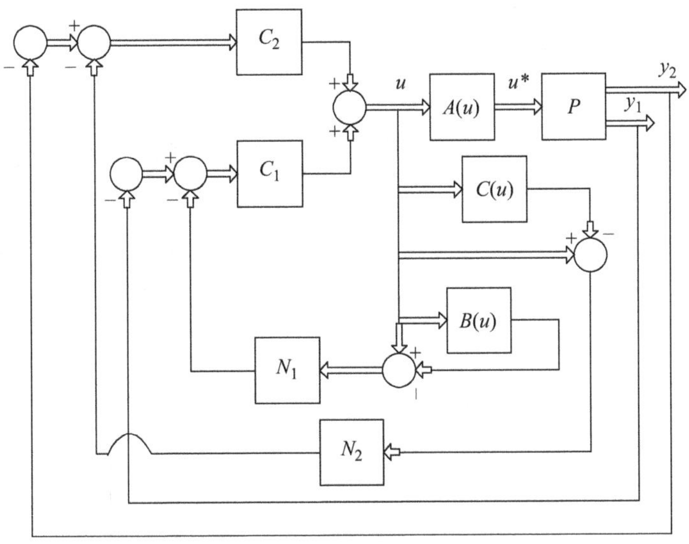

Figure 7.20 SITO feedback system with NDC

the NDC around C_1 is active, and the equivalent feedback diagram is shown in Figure 7.21. The linear system in feedback connection to A is as follows:

$$T_{eq}(s) = -\frac{u}{u^*} = \frac{P_2(s)C_2(s) + P_1(s)C_1(s) - C_1(s)N_1(s)}{1 + C_1(s)N_1(s)} \tag{7.9}$$

Similarly, if $u \in u_2$, then the system is equivalent to that shown in Figure 7.22 and the equivalent linear system is

$$T_{eq}(s) = -\frac{u}{u^*} = \frac{P_1(s)C_1(s) + P_2(s)C_2(s) - C_2(s)N_2(s)}{1 + C_2(s)N_2(s)} \tag{7.10}$$

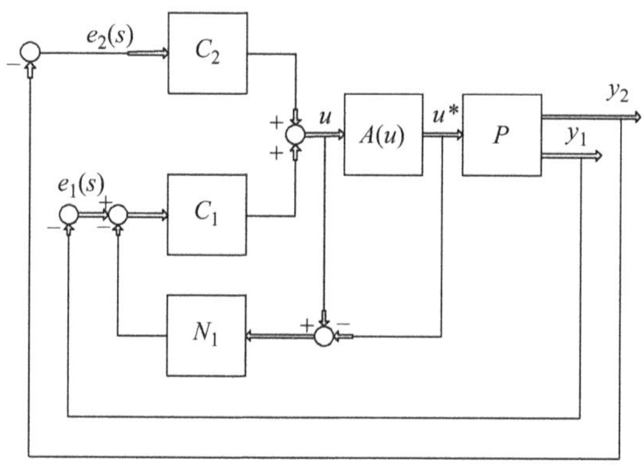

Figure 7.21 Equivalent system when $u \in u_1$

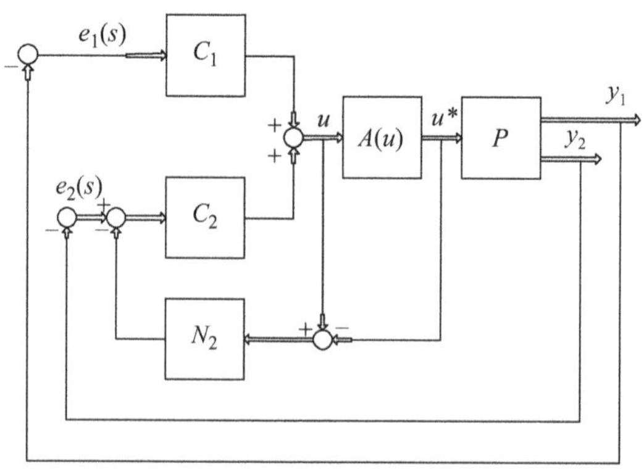

Figure 7.22 Equivalent system when $u \in u_2$

Often, one of the two compensator functions of the SITO controller limits the sector width for which the system is absolutely stable. Consider the case where the compensator $C_1(s)$ is very aggressive (large loop gain, steep roll-off that results in the Nyquist locus to be far in the second and third quadrants) and $C_2(s)$ is less aggressive (relatively small loop gain and a shallow roll-off that results in the Nyquist locus residing primarily in the first and fourth quadrants). In this case, the NDC need only be applied to the C_1 as shown in Figure 7.23, which considers position and rate nonlinearities of the actuator. M in this case would be a derivative operation on the actuator input and the second NDC path corrects the system in actuator rate saturation. Note that the effects of the saturations must be considered separately.

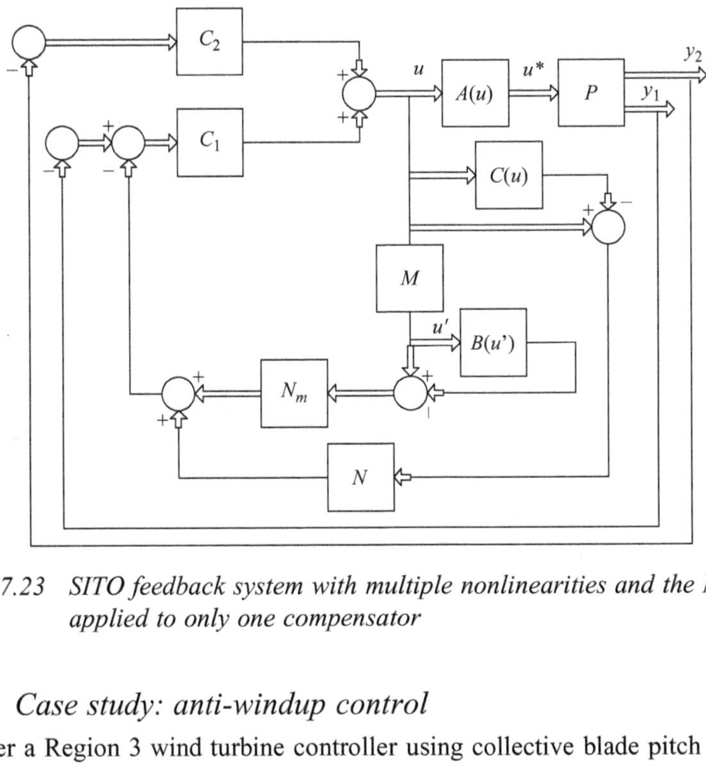

Figure 7.23 SITO feedback system with multiple nonlinearities and the NDC applied to only one compensator

7.3.1 Case study: anti-windup control

Consider a Region 3 wind turbine controller using collective blade pitch for rotor rate regulation. The plant is nonlinear, so the model is obtained by linearizing the system dynamics at a particular operating point related to wind speed. A type-1 controller is designed to reject step wind changes. The expected integrator windup is observed when the blade pitch reaches its limit. An anti-windup compensator is applied and this effect is reduced. When severe turbulent conditions are simulated, poor regulation performance is observed, and the collective pitch rate is exceeded over significant intervals (Figure 7.24). It is noted that this is associated with a collective blade pitch *rate* saturation. A second anti-windup loop is placed in parallel to the first (Figure 7.25). The nonlinear block is a deadzone with a width

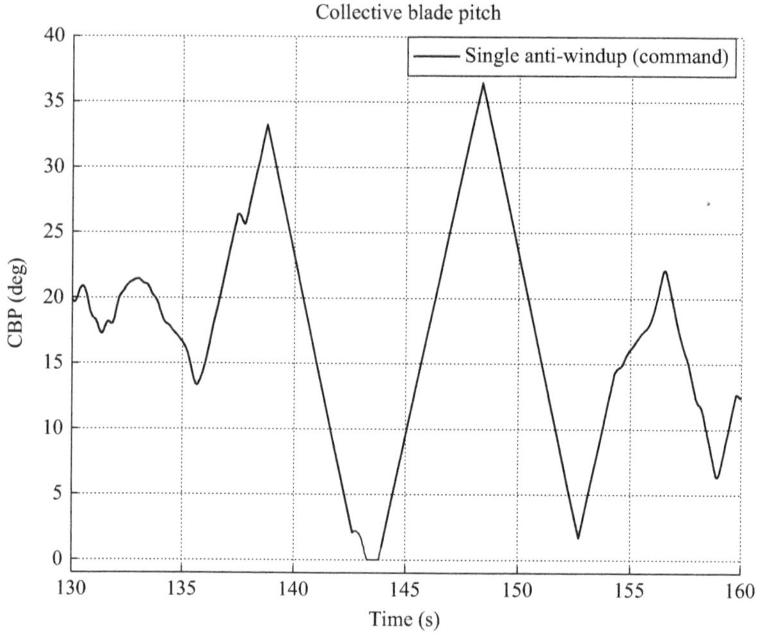

Figure 7.24 Wind turbine collective blade pitch with one anti-windup loop

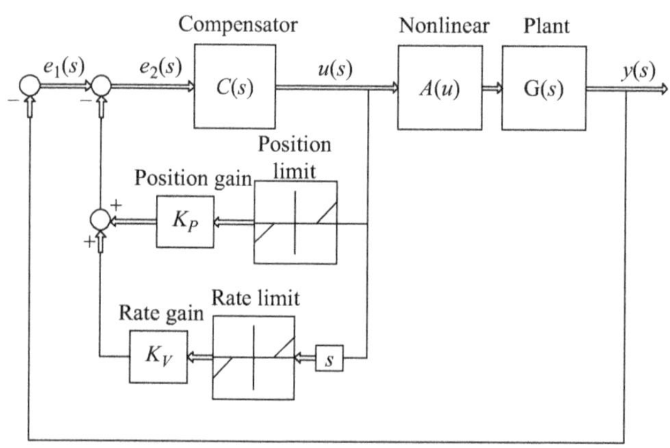

Figure 7.25 Multiple-path anti-windup control block diagram

equal to the blade pitch rate limits. The test is run again with the same wind conditions, and the blade rate saturation is reduced (Figure 7.26). The high-speed shaft rate is shown in Figure 7.27; it is noted the dual anti-windup system has superior performance. The effectiveness of the gain reduction of the anti-windup controller indicates that the relative stability of the controller with actuator rate saturation is insufficient and the loop gain reduction provided by the second anti-windup loop ameliorates this.

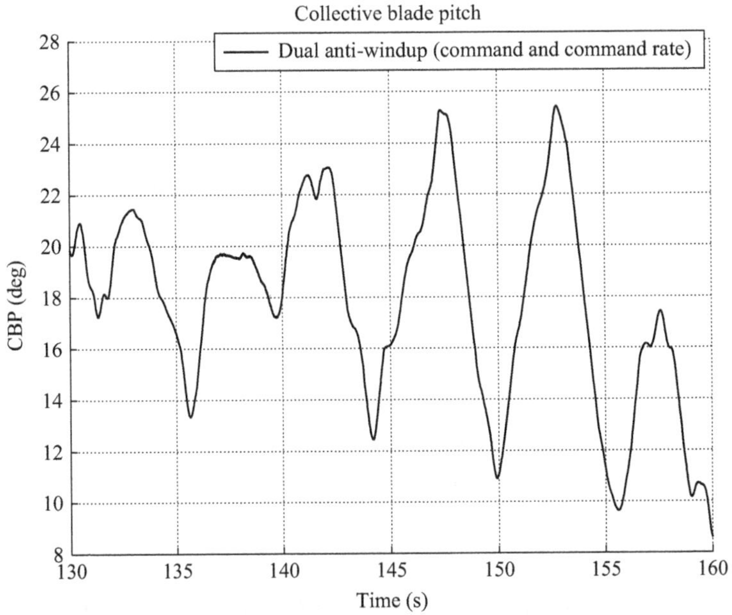

Figure 7.26 Wind turbine collective blade pitch with two anti-windup loops

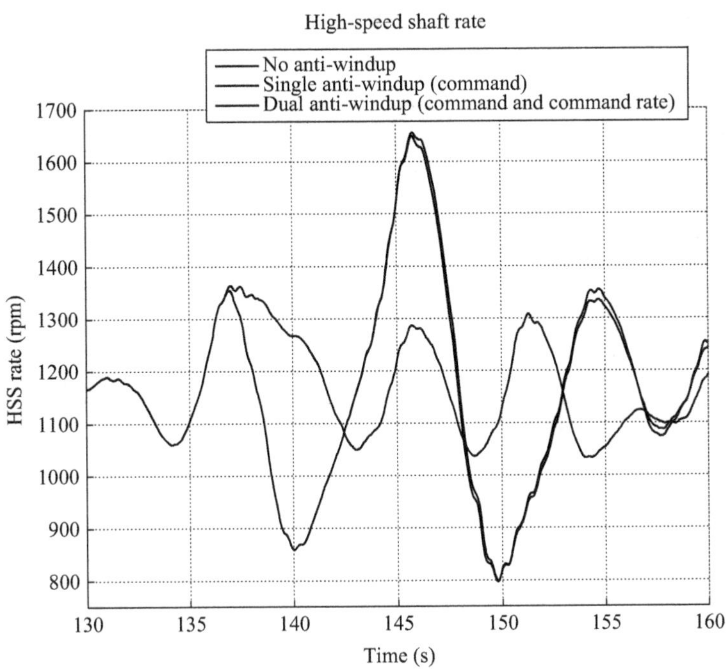

Figure 7.27 High-speed shaft rate

7.3.2 Case study: nonlinear dynamic compensation for multiple saturations

The high-order controller with NDC applied to the rudder roll stabilization controller discussed in Chapter 6 is AS only if the rudder is not rate saturated. Unfortunately, rate saturation is a key limitation in this application, especially for rudder steering machines on larger vessels. A new scheme is presented in this section that provides AS for rudder angle and rudder rate saturation. A block diagram of the controller is shown in Figure 7.28. The saturation links in the NDC called 'rate loop' and 'position loop' are identical to the saturations 'rudder rate limiter' and 'rudder limiter,' respectively, in the rudder model. In rudder rate saturation (no angle saturation), the equivalent compensator is shown in Figure 7.29, which allows AS analysis described in Chapter 5. The equivalent linear system connected to the saturation nonlinearity is

$$T_{e_r}(s) = \frac{1 + G_r(s)C_r(s) + G_y(s)C_y(s) - C_r(s)C_{n_1}(s)\left(\frac{s^2}{s+1}\right)}{s\left(1 + C_r(s)C_{n_1}(s)\left(\frac{s}{s+1}\right)\right)} \tag{7.11}$$

Transfer function C_{n_1} is chosen such that $T_{e_r} = T_e$, and thus the system is AS for the rudder rate saturation.

In rudder angle saturation (no rate saturation), the equivalent compensator is shown in Figure 7.30. The saturation limits are identical to the rudder angle limits. This system connected to the plant yields the feedback connection to the saturation nonlinearity, and AS analysis is possible.

$$T_{e_a}(s) = \frac{P_r(s)C_r(s) + P_y(s)C_y(s) - N_c(s)C_r(s)}{1 + N_c(s)C_r(s)} \tag{7.12}$$

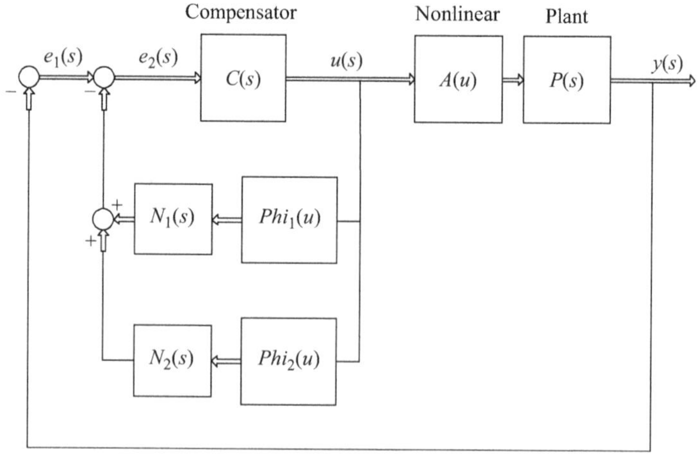

Figure 7.28 Heading control and RRS with multiple feedback path NDC

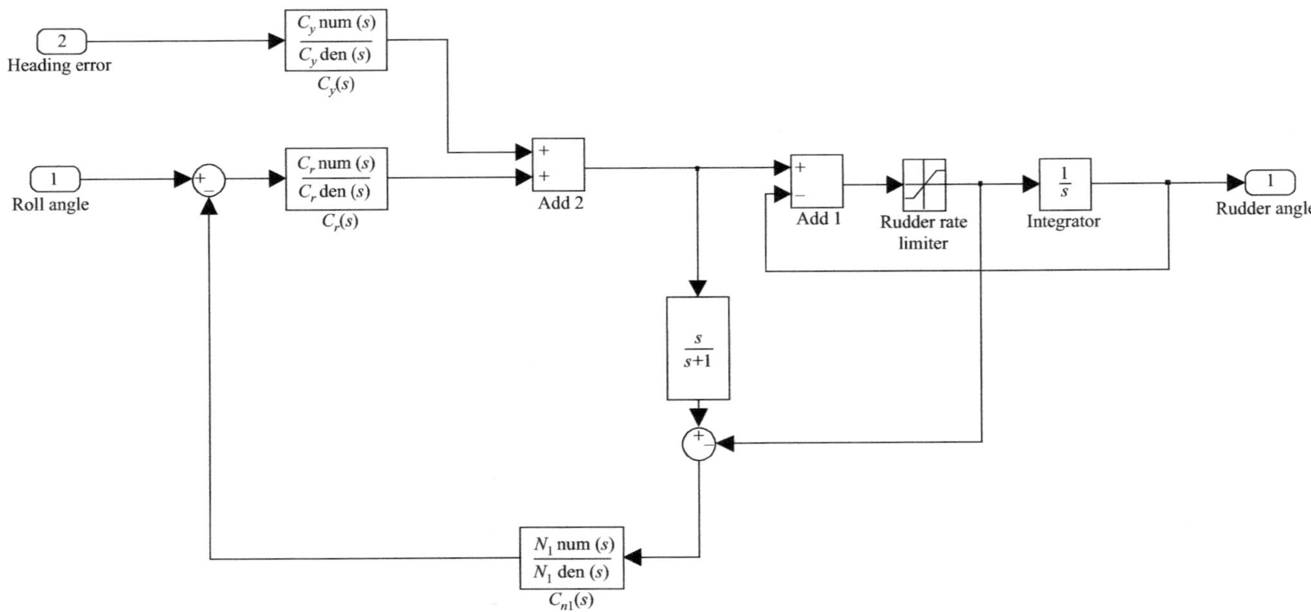

Figure 7.29 Equivalent RRS compensator in the rate saturation condition

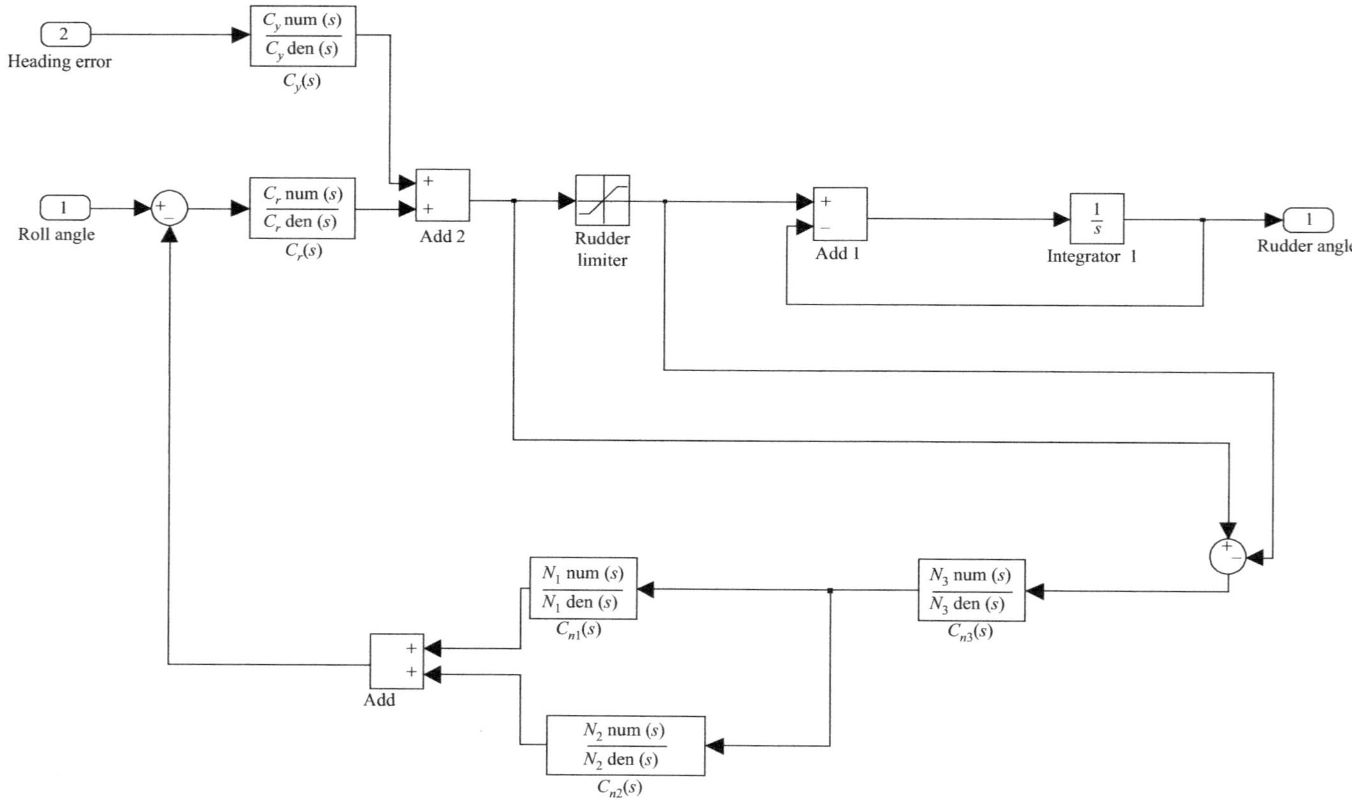

Figure 7.30 Equivalent RRS compensator in angle saturation condition

where $P_r(s) = G_r(s)\frac{1}{s+1}, P_y(s) = G_y(s)\frac{1}{s+1}$, and $N_c = C_{n3}(C_{n_1} + C_{n_2}) = C_n$. The structure of N_c is chosen because nonminimum phase zeros in C_{n_1} make the filter $\frac{C_n}{C_{n1}}$ unstable, thus a cascade of two filters is not feasible. With the selected N_c, $T_{e_a} = T_e$ and the system is AS for the rudder angle saturation.

With this multipath NDC, the high-performance Nyquist-stable rudder roll stabilizer is AS for angle and rate saturations. The last stability consideration is the condition of *simultaneous* angle and rate saturation. This case does not lend itself to AS analysis. As the Popov Criterion is only a sufficient condition, this is not an indication of instability. Figure 7.31 shows the Nyquist plot for the very large signal loop gain for the multipath NDC RRS with heading control around the rudder actuator. This is an approximation of the loop gain when the saturation paths in the NDC shown in Figure 7.28 are negligible compared to the linear paths (the actuator is saturated in both rate and angle with large input signals). The critical point is not encircled if the plot collapses on the origin due to broadband saturation. No instabilities have been experienced in simulations where simultaneous saturations occur. While not rigorously proven, the results indicate stability retention in the simultaneous saturation condition.

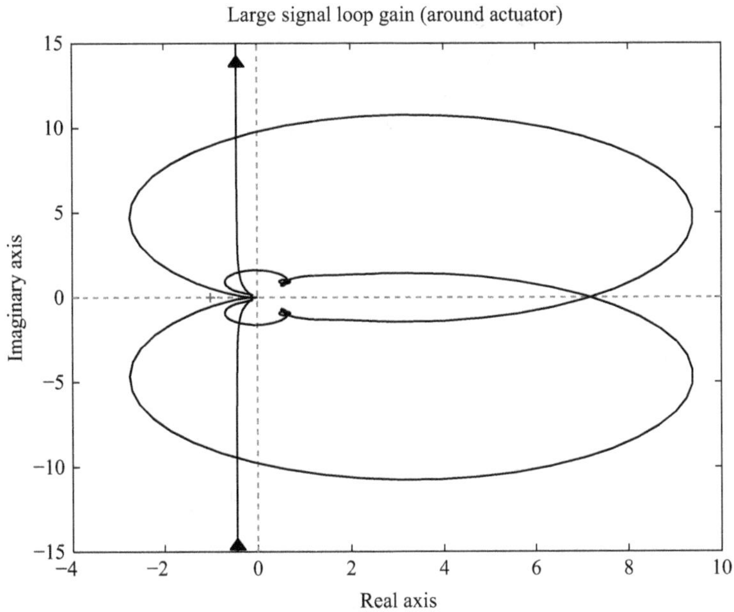

Figure 7.31 Very large signal loop gain

The efficacy of the roll stabilizer is shown in computer simulations. Three RRS will be compared: proportional derivative (PD), Nyquist-stable control with angle saturation NDC and Nyquist-stable control with multipath NDC. The PD heading controller described previously is used in conjunction with all three RRS systems. Three rudder rate limits are considered: 20, 15 and 10 deg/s. Pierson-Moskowitz linear approximation model using $\sigma_r = 4.0$, $\sigma_y = 1$, $\omega_0 = 0.5$ rad/s and $\zeta = 0.1$ is

Table 7.1 Roll reduction percentage ('x' indicates immediate oscillation)

	Rudder rate		
	20 deg/s	**15 deg/s**	**10 deg/s**
PD	68.0	67.5	65.4
HO+NDC	88.5	47.1	x
HO+multipath NDC	86.7	83.8	71.5

used to model wave disturbances. A quantitative measure of the relative performance is provided by the roll reduction percentage defined by Oda *et al.* [74].

$$\text{Roll reduction} = \frac{\text{AP} - \text{RRS}}{\text{AP}} \times 100 \tag{7.13}$$

where AP is the standard deviation of roll rate with the heading controller on, RRS off and RRS the standard deviation of roll rate with both the heading and RRS on. Table 7.1 shows the roll reductions for the PD controller, high-order controller with rudder angle NDC only (HO+NDC) and high-order controller with multipath NDC (HO+multipath NDC).

The multipath NDC system provides superior performance down to 10 deg/s with the exception of a slight inferiority to HO+NDC with the fastest rudder. Figure 7.32 shows the roll angles in open-loop, closed PD roll stabilizer, closed HO+NDC and closed HO+multipath NDC with the 15 deg/s rudder. The heading controller is closed in all three cases, with a heading command of 0. The yaw angle with the multipath NDC controller open and closed loop is shown in

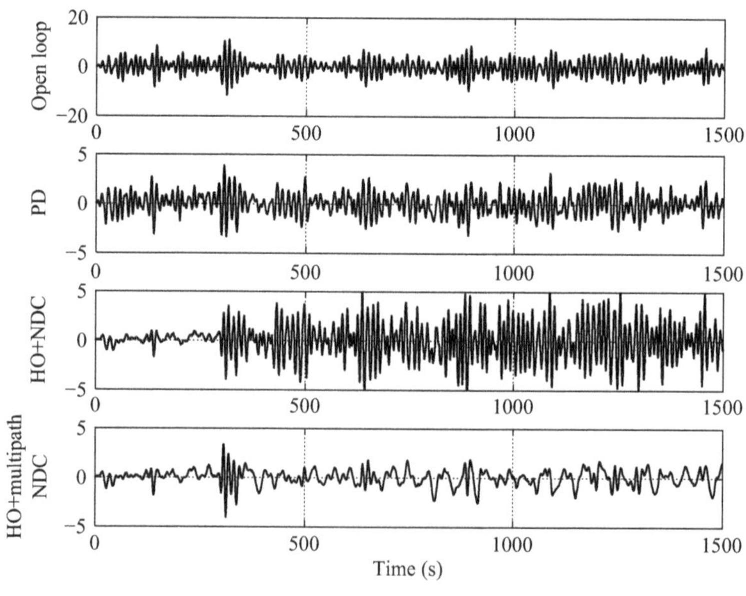

Figure 7.32 Roll angles

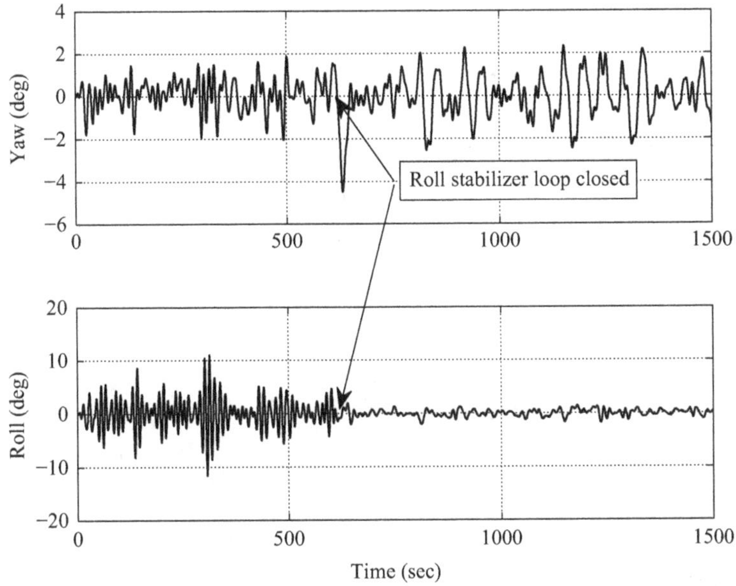

Figure 7.33 Yaw angle with multipath NDC open and closed

Figure 7.33. The disturbance event at approximately 310 s causes oscillation in the high-order controller with rudder angle NDC only, resulting in poor roll reduction performance. Figure 7.34 shows the rudder angle and rates for this controller, where the rudder oscillations are evident. This represents a serious limitation of this control system when slower steering machines are implemented. The rudder angle

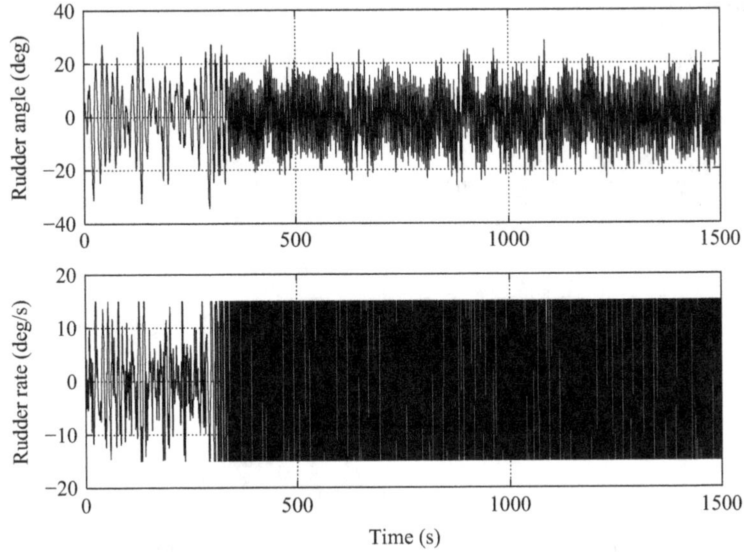

Figure 7.34 Rudder angle/rate with closed-loop angle only NDC

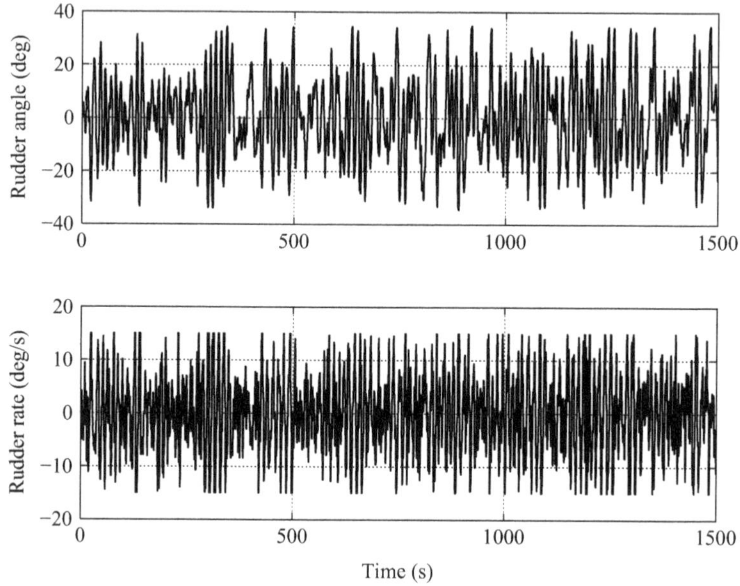

Figure 7.35 Rudder angle/rate with closed-loop multipath NDC

and rate signals for the same conditions are shown for the multipath NDC controller in Figure 7.35. It is noted that in contrast to the high-order controller with angle only NDC, this system does not oscillate. The angle and rate summing junction outputs are shown in Figure 7.36 (15 deg/s rudder). There is saturation when these

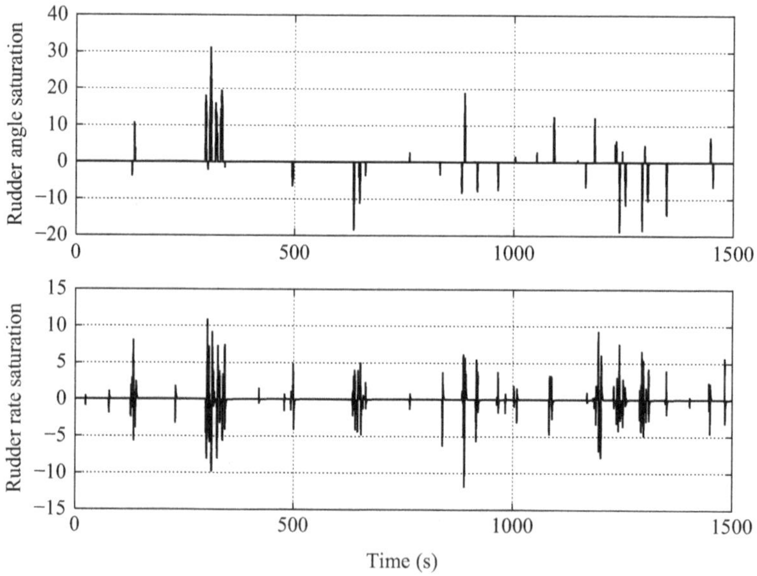

Figure 7.36 Multipath NDC deadzone outputs

functions have nonzero values. The figure shows that both rudder angle and rate saturations occur over the 1500 s time interval (Table 7.1).

7.4 Variable gain for SITO feedback systems

The focus of the book thus far is primarily on the control of LTI SISO systems. Taking into consideration the feedback limitations presented by the plant, actuators, sensors, environment, etc. the implementation of feedback maximization can produce a high-performance controller for this type of system. Unfortunately, rarely is the control designer presented with such a tidy problem. More often than not, the plant will not satisfy the conditions of linearity, or time invariance. This ubiquitous problem of actuator saturation has already been addressed in the previous chapter. It is very common for the plant dynamics to be a function of environmental states. Aircraft autopilots are typically designed using linear control theory; however, aircraft dynamics change as a function of its location in the 'flight envelope', the aircraft's airspeed, and altitude. This is also the case with surface vessel autopilots, where the dynamics are a function of the ship's speed. Wind turbines are nonlinear systems; however, at particular wind speeds their dynamics can be described using linear models.

An interesting application for gain scheduling is a variable gain swap for a SITO controller. The block diagram of this system is shown in Figure 7.37. Recall the concept of frequency separation for these controllers, whereby disjoint frequency intervals are defined where the feedback for each of the two plant outputs is negative. Given the plant transfer function vector $P(s) \in \mathbf{F}^{2\times1}(s)$, the outputs are rated according to importance (e.g. regulation of $p_1(s)$ is of greater importance than $p_2(s)$). The nominal feedback controller $C(s) \in \mathbf{F}^{1\times2}(s)$ is designed. A system is then designed that 'swaps' gain from the 'more important' controller to the 'less

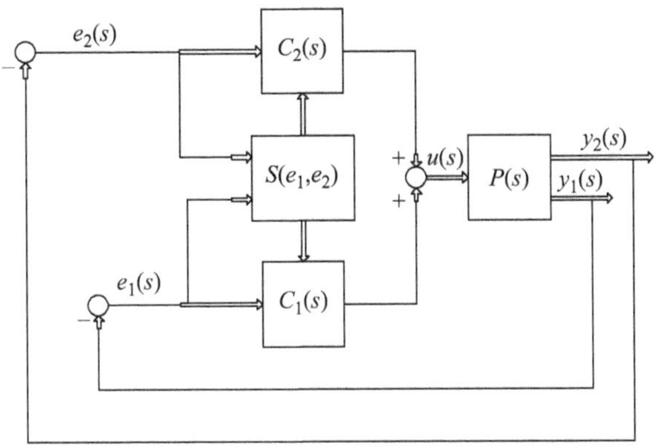

Figure 7.37 SITO Controller with Gain Swapping

important' compensator $C(s)$ as a function of the output of the more important plant. This improves the performance of the important plant without requiring greater overall feedback (it is assumed that the bandwidth is as high as the system limitations allow). It is imperative that the relative stability of the system is sufficient at all possible values of the output of $S(e_1, e_2)$. This typically requires wider frequency intervals of minimum phase margin than what is required for the fixed SITO controller. The speed of the variable gain system should be slower than the dynamics of the plant. A low pass filter at the output of the variable gain system can effectively 'slow' the gain changes.

7.4.1 Case study: HSS rate error variable gain SITO controller

The requisite frequency separation for the SITO collective blade pitch controller described in Chapter 6 is a substantial limitation. As rotor rate regulation is the primary function of the collective blade pitch controller in Region 3, a variable gain strategy is proposed to apply more rate feedback when error is large. This comes at the expense of lowering tower acceleration feedback to allow increased rate feedback bandwidth.

To quantify the relative performance of the SITO controllers, two dimensionless parameters are defined. The high-speed shaft rate standard deviation to rated speed ratio (SDRS) is

$$\text{SDRS} = \frac{\sigma_{hss}}{\omega_{hss(rated)}} \tag{7.14}$$

where σ_{hss} and $\omega_{hss(rated)}$ are the high-speed shaft rate standard deviation and the Region 3 rated high-speed shaft rate, respectively, in the same units. As the turbines in this study are constant torque, SDRS provides a measure of power variation in Region 3.

The secondary goal of the collective blade pitch SITO controller is tower fore/aft acceleration attenuation. The tower fore/aft acceleration ratio (TFAAR) is

$$\text{TFAAR} = \frac{\sigma_{a(cl)}}{\sigma_{a(ol)}} \tag{7.15}$$

where $\sigma_{a(cl)}$ is the tower fore/aft acceleration standard deviation with the SITO controller closed, and $\sigma_{a(ol)}$ is the tower fore/aft acceleration standard deviation with high-speed shaft rate feedback only (referred to as the *baseline*).

FAST simulations are performed for both turbines over 10 min in turbulent wind. Table 7.2 shows the rated high-speed shaft rate, the high-speed shaft rate standard deviation and tower fore/aft acceleration in open- and closed-loop conditions for both turbines with the SITO controllers with loop transmissions shown in Figure 7.38.

Table 7.2 SITO control performance in turbulent wind

	CART2	5 MW
Rated shaft speed (rpm)	1800	1174
HSS rate std (rpm)	31.9	83.46
Tower fore/aft acc std (ol) (m/s/s)	0.3549	0.2988
Tower fore/aft acc std (cl) (m/s/s)	0.2193	0.1719

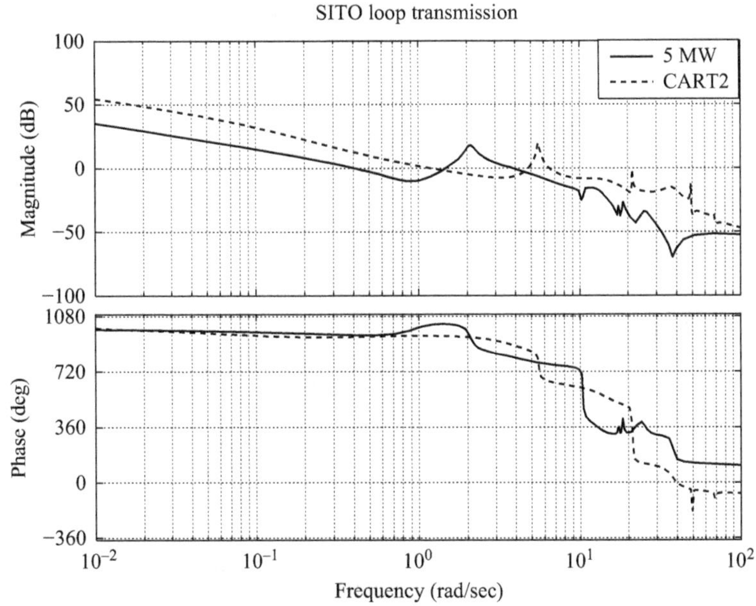

Figure 7.38 Type 1 Region 3 collective blade pitch SITO loop transmissions for the CART2 (dotted line) and 5 MW (solid line) wind turbines at 18 m/s

The TFAAR for the CART2 and 5 MW turbines are 0.62 and 0.58, respectively. The performance is comparable, which is consistent with the fact that both controllers apply close to the same feedback at the tower first mode frequency. The SDRS for the CART2 and 5 MW turbines are 0.018 and 0.071, respectively. The 5 MW SDRS is four times greater than that of CART2. This is consequence of the 5 MW wind turbine plant dynamics restricting available feedback to greater effect than for the CART2 turbine.

The variable gain algorithm is a piecewise continuous function of high-speed shaft rate $(\omega_{hss}) = 1174$ rpm and gear ratio $r_g = 97$. The output is a gain applied to the nominal tower acceleration feedback compensator, $k_{struct(5v)}(\omega_{hss})$ and the nominal high-speed shaft rate feedback compensator, $k_{rate(5v)}(\omega_{hss})$. Variable gain $k_{struct(5v)}(\omega_{hss})$ is defined in Table 7.3. The constants $\omega_{hss(i)}$, $i = 1, 2, 3, 4$, are threshold values. If the high-speed shaft rate is in the interval bounded by $\omega_{hss(2)}$ and $\omega_{hss(3)}$, the gain $k_{struct(5v)}(\omega_{hss})$ is unity and the collective blade pitch controller

Table 7.3 HSS rate error variable gain algorithm

ω_{hss}	$k_{struct(5v)}(\omega_{hss})$
$\omega_{hss} \leq \omega_{hss(1)}$	0.05
$\omega_{hss(1)} < \omega_{hss} \leq \omega_{hss(2)}$	$\frac{1}{r_g}(\omega_{hss} - \omega_{hss(1)})$
$\omega_{hss(2)} < \omega_{hss} \leq \omega_{hss(3)}$	1.0
$\omega_{hss(3)} < \omega_{hss} \leq \omega_{hss(4)}$	$-\frac{1}{r_g}(\omega_{hss} - \omega_{hss(3)}) + 1$
$\omega_{hss} > \omega_{hss(4)}$	0.05

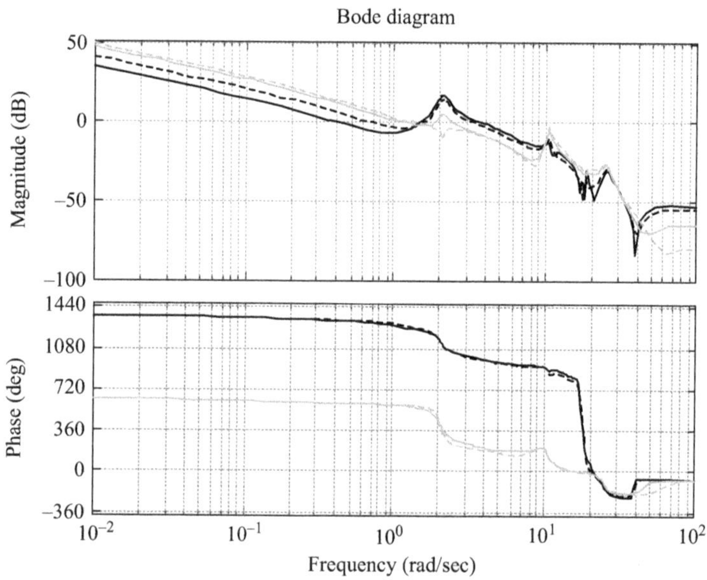

Figure 7.39 Variable gain 5 MW SITO loop transmission. Dark solid line is the standard SITO loop transmission. Dark dotted line, grey solid line and grey dotted line loop transmission are in conditions of increasing high-speed shaft rate error

is the SITO controller previously described. If the shaft speed is outside this interval, $k_{struct(5v)}(\omega_{hss})$ is reduced linearly until the shaft speed is outside the interval bounded by $\omega_{hss(1)}$ and $\omega_{hss(4)}$ where this gain is fixed at 0.05. Given this gain, the rate regulator gain is $k_{rate(5v)}(\omega_{hss}) = -4k_{struct(5v)}(\omega_{hss}) + 5$.

For the 5 MW wind turbine, the threshold values are $\omega_{hss(i)}$ are 1039, 1125, 1222 and 1309 rpm for $i = 1, 2, 3, 4$. Figure 7.39 shows the loop transmission of the variable gain controller with different high-speed shaft rate errors. The loop transmission associated with the largest rate error (light blue) has negligible feedback at the first tower mode frequency (approximately 2 rad/s) and has about one octave more bandwidth (1 rad/s) than the nominal SITO controller (0.4 rad/s).

Figures 7.40 and 7.41 show the high-speed shaft rate for the 5 MW turbine with high-speed shaft rate feedback only (baseline), fixed gain SITO and variable

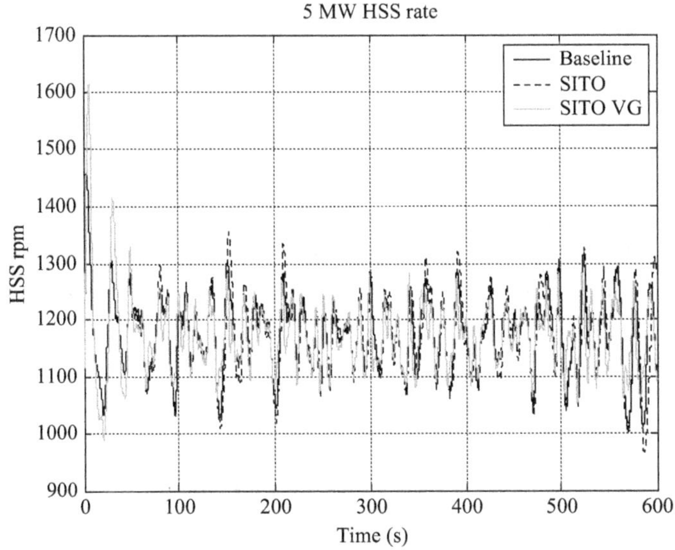

Figure 7.40 5MW high-speed shaft rate for baseline (solid black line), fixed gain SITO (solid dotted line) and variable gain SITO (grey solid line) collective blade pitch control

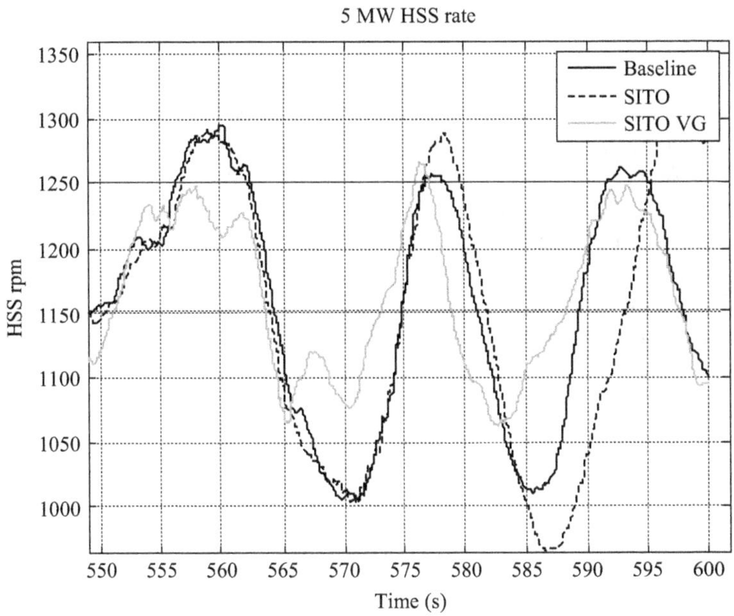

Figure 7.41 5MW high-speed shaft rate for baseline, fixed gain and variable gain collective blade pitch control. 1250 rpm is the variable gain threshold value

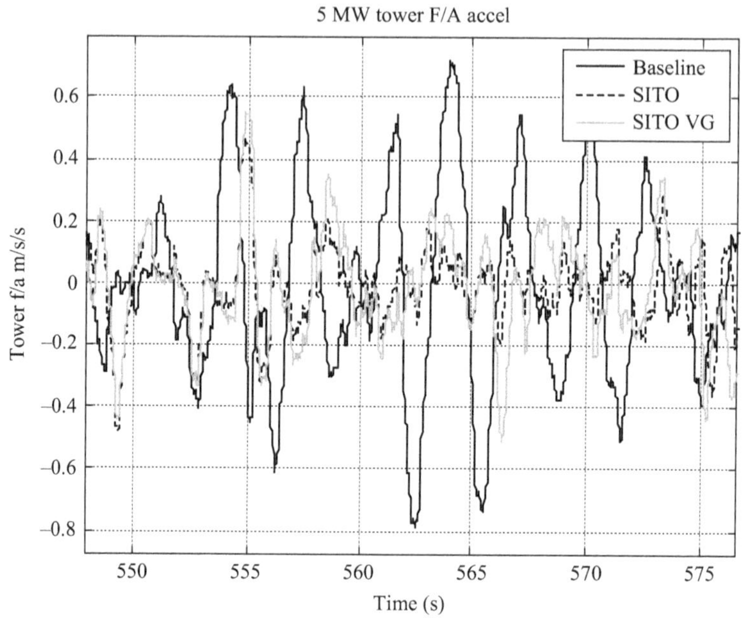

Figure 7.42 5MW tower fore/aft acceleration for baseline, fixed gain and variable gain collective blade pitch control

gain SITO control in turbulent wind conditions. It is evident that variable gain system has superior performance compared to the fixed gain SITO controller. The SDRS for the variable gain controller is 0.0575, a 19% improvement over the fixed SITO controller. Figure 7.42 shows the tower fore/aft acceleration for the same controllers. As the acceleration feedback is reduced as a function of rate error, the variable gain controller has inferior performance to the fixed gain SITO system for tower load mitigation, but clearly superior to the baseline controller. The TFAAR for the variable gain controller is 0.67, a 16% reduction in performance compared to the fixed gain controller. This relative performance is a consequence of the variable gain strategy reducing the tower mode feedback when shaft rate error exceeds the threshold.

7.5 Exercises

1. Explain why a Nyquist-stable control should not be deployed without non-linear dynamic compensation.
2. Design a PI controller for the plant $P(s) = \frac{1000}{s+10s+100}$. Write a Simulink model with a unit saturation in the forward path. The reference is 100 1^+. Design an anti-windup compensator for this controller. Compare the step response of these two systems.

3. Replace the saturation with the multiple saturation in the RRS case study (position and velocity saturation). Design a dual anti-windup controller for this system.
4. How can the control designer approximate the performance of the feedback system with nonlinear dynamic compensation in the very large signal condition?
5. Design a nonlinear dynamic compensator for the controller designed in Chapter 6, Exercise 6.
6. Design a Nyquist-stable controller for the plant of Exercise 12 in Chapter 6. There is a unit saturation in the forward path between the compensators and the plant. Design an NDC for the $C_2(s)$ controller. Show that the system is absolutely stable.
7. Design a variable gain system for the controller of the previous problem. Write a Simulink model to test the efficacy of the controller.

Chapter 8

References

The material in Chapter 2 is found in many publications; References 1 and 2 are good ones. The treatment of state space models follows References 3 and 4. The definitions and concepts in Chapter 3 follow References 10, 11 and 12. The generalized solution of sensitivity integral (the proof is in the appendix) originally reported by Bode is from Reference 16. Descriptions of feedback limitations are readily found; References 15 and 20 are good. Chapter 4 is from Lurie (Reference 10). The material in Chapter 5 is found in many publications. Definitions of the stability of linear systems are found in Chen [3], Skogestad and Postlewaite [33] and Jiguan Lin Varaiya [34]. The discussion of hidden modes, internal stability and the generalized Nyquist Criterion follows References 39 and 41. Gershgorin's Theorem is presented in Reference 42 and the case study utilizing it is from Reference 65. The material on Lyapunov and absolute stability is from Khalil [44], Murray *et al.* [45] and Slotine and Weiping [46]. The robot set point control case study is from Reference 50. The material in Chapter 6 follows References 10 and 12. The case study quantifying feedback for wind turbine control is from Reference 66. The material on Nyquist-stable systems and nested actuation is rooted in the work by Lurie and O'Brien [54–60]. The material on nonlinear dynamic compensation in Chapter 7 is from Lurie (e.g. References 10, 11, 58). The design and stability analysis of a Nyquist-stable controller with nonlinear dynamic compensation applied to a vibration suppression system is from Carruthers and O'Brien [82]. Variable gain and nonlinear dynamic compensation applied to SITO plants and for systems with multiple nonlinearities is from References 63, 66 and 81.

Appendix

Proof of Bode sensitivity integral

Consider the response of a feedback system with sensor noise, and a disturbance additive at the output. The response of the system to the disturbance is $y_d(s) = S(s)d(s)$, where $S(s)$ is the sensitivity function of the feedback system and $y_d(s)$ is the Laplace transform of the system's response to disturbance signal $d(s)$. The response of the system to the sensor noise, $n(s)$, is $y_n(s) = -M(s)n(s)$, where $M(s)$ is the complementary sensitivity function. Assume that the loop transmission function $T(s)$ has no hidden unstable modes. Then the feedback system is stable if $S(s)$ is bounded in the CRHP, and CRHP poles and zeros of the plant $P(s)$ and compensator $C(s)$ appear with the same multiplicity in $T(s)$.

It is assumed that $T(s)$ can be factored thusly.

$$T(s) = \tilde{T}(s)B_p^{-1}(s)B_z(s) \tag{A.1}$$

where $B_z(s)$ is the *Blaschke* product of N_z ORHP zeros including multiplicities, $Z = \{z_i : Re(z_i) > 0;\ T(z_i) = 0\}$

$$B_z(s) = \prod_{i=1}^{N_z} \frac{z_i - s}{\overline{z_i} + s} \tag{A.2}$$

Note that B_z is an all-pass, unit gain system. Similarly,

$$B_p = \prod_{i=1}^{N_p} \frac{p_i - s}{\overline{p_i} + s} \tag{A.3}$$

is the Blaschke product of N_p ORHP poles including multiplicities, $P = \{p_i : Re(p_i) > 0, T^{-1}(p_i) = 0\}$. $\tilde{T}(s)$ is proper with no poles or zeros in the ORHP. For each CRHP zero z of multiplicity m,

$$\begin{aligned}
S(z) &= 1 \\
\frac{d}{ds} S\,|_{s=z} &= 0 \\
\vdots &= 0 \\
\frac{d^i}{ds^i} S\,|_{s=z} &= 0
\end{aligned}$$

for $i = 1, 2, \ldots, m-1$ and S is the sensitivity function. Similarly, at each CRHP pole p of multiplicity n,

$$M(p) \sim 1$$
$$\frac{d}{ds}M|_{s=p} = 0$$
$$\vdots = 0$$
$$\frac{d^i}{ds^i}M|_{s=p} = 0$$

for $i = 1, 2, \ldots, n-1$ and M is the complimentary sensitivity function. Assume the feedback system is stable. Then S and M have no CRHP poles. Express the sensitivity function and its derivative constraints in terms of the magnitude $|S(j\omega)|$ by removing the zeros of S at the ORHP poles of T. Factor the sensitivity function using the Blaschke product.

$$S(s) = \tilde{S}(s)B_p(s) \tag{A.4}$$

Since $|B_p(j\omega)| = 1\ \forall\ \omega, |S(j\omega)| = |\tilde{S}(j\omega)|\ \forall\ \omega$, it is necessary to constrain $\log S(s)$ and $\frac{d^i}{ds^i}\log S(s)$ at infinity. Given function $F(s)$, consider the following class of functions:

$$N(R) = \sup_{\theta} |F(Re^{j\theta})| \tag{A.5}$$

where $\theta \in \left[-\frac{\pi}{2}, \frac{\pi}{2}\right]$. $F(s)$ is said to be in class R provided $lim_{R\to\infty}\frac{1}{R}N(R) = 0$. If $T(s)$ is a proper rational function, then $\log T(s)$ and $\frac{d^i}{ds^i}\log T(s)$ are in class R. Constraints of $S(s)$ at ORHP zeros can be expressed in terms of $S(j\omega)$.

Theorem Let $f(s)$ be analytic and nonzero in the CRHP except for possible zeros on the imaginary axis. Assume $\frac{d^i}{ds^i}\log f(s)$ is in class R for $i = 0, 1, \ldots$. Then at each point $s_0 = x_0 + jy_0$, $x_0 > 0$.

$$\log|f(s_0)| = \frac{1}{\pi}\int_{-\infty}^{\infty} \log|f(j\omega)| \frac{x_0}{x_0^2 + (y_0-\omega)^2} d\omega$$

$$\frac{d^i}{ds^i}\log f(s)_{s=s_0} = \frac{1}{\pi}\int_{-\infty}^{\infty} \frac{d^i}{ds^i}\log f(s)_{s=j\omega} \frac{x_0}{x_0^2 + (y_0-\omega)^2} d\omega$$

The proof is an example of the solution of Poisson integrals [16]. Set $f(s) = \tilde{S}(s)$. Recall $S(s) = \tilde{S}(s)B_p(s)$ and note $S(z) = \tilde{S}(z)B_p(z) = 1$. So

$$\log|\tilde{S}(z)| = \frac{1}{\pi}\int_{-\infty}^{\infty} \log|S(j\omega)| \frac{x_0}{x_0^2 + (y_0-\omega)^2} d\omega \tag{A.6}$$

can be expressed as the following.

Theorem Let $z=x+jy$ be an ORHP zero of $T(s)$. Then if the feedback system is stable, the sensitivity function must satisfy the following:

$$\pi\log|B_p^{-1}(z)| = \int_{-\infty}^{\infty} \log|S(j\omega)|d\theta_z(\omega) \tag{A.7}$$

where $\theta_z(\omega)=\arctan\frac{\omega-y}{x}$.

A.1 Bode's integral theorem

The bandwidth constraint is quantified by the following inequality:

$$|T(j\omega)| \le \frac{A}{\omega^{1+k}} \tag{A.8}$$

for $\omega > \omega_0$. $k > 0$, so the loop transmission roll-off is greater than first order. This assumption is not restrictive, as most plant frequency responses roll off at high frequency at a greater slope than this. An upper bound on loop transmission magnitude at some frequency ω_0 is assumed.

$$\frac{A}{\omega_0^{1+k}} \le a \tag{A.9}$$

Theorem Assume loop transmission $T(s)$ has a finite number of ORHP poles $P=\{p_i : Re(p_i) > 0,\ T^{-1}(p_i)=0\}$ for $i=1, 2, \ldots N_p$ including multiplicities and assume $lim_{R\to\infty}sup_{|s|\ge R, Re[s]\ge 0}R|T(s)|=0$. Then, if the closed-loop system is stable, the sensitivity $S(s)$ must satisfy

$$\pi\sum_{i=1}^{N_p} Re[p_i] = \int_0^{\infty} \log|S(j\omega)|d\omega \tag{A.10}$$

Proof Recall $\log|f(s_0)| = \frac{1}{\pi}\int_{-\infty}^{\infty} \log|f(j\omega)|\frac{x_0}{x_0^2+(y_0-\omega)^2}d\omega$, where $f(s)$ is analytic and nonzero in the CRHP except for possible $j\omega$ axis zeros. $f(s)=\tilde{S}(s), s=x>0$, so $\log|\tilde{S}(x)| = \frac{2}{\pi}\int_0^{\infty} \log|S(j\omega)|\frac{x}{x^2+(y-\omega)^2}d\omega$. Note that $\log|S(j\omega)|\frac{x^2}{x^2+(y-\omega)^2}$ converges to $\log|S(x)|$ as $x\to\infty$, which suggests $lim_{x\to\infty}x\log|\tilde{S}(x)| = limx\to\infty\frac{2}{\pi}\int_0^{\infty} \log|S(j\omega)|\frac{x^2}{x^2+(y-\omega)}\,d\omega$ can be used to evaluate the integral of the log sensitivity frequency response over all nonnegative frequencies. Consider the following.

1. $\int_0^{\infty} \log|S(j\omega)|d\omega \simeq \int_0^{\overline{\omega}} \log|S(j\omega)|d\omega$ for sufficiently large $\overline{\omega}$.
2. For $\omega\in[0,\overline{\omega}]$, the sequence $\frac{n^2}{n^2+\omega^2}$ converges to 1 as $n\to\infty$.

The power series expansion of $\log S(s) = -\log[1 + T(s)]$ for $|T(s)| < 1$ is

$$\log S(s) = -T(s) + \frac{T^2(s)}{2} + HOT \tag{A.11}$$

Recall the assumption $lim_{R\to\infty} sup_{|s|\geq R, Re[s]\geq 0} R|T(s)| = 0$, so $\omega \log(S(j\omega)) \to 0$ as $\omega \to \infty$, so there exists a frequency ω_0 and positive constants M_0 and δ such that $|\log|S(j\omega)|| \leq \frac{M_0}{\omega^{1+\delta}}$ for $\omega > \omega_0$. For $\omega_x > \omega_0$

$$\int_{\omega_x}^{\infty} |\log|S(j\omega)||d\omega \leq \int_{\omega_x}^{\infty} \frac{M_0}{\omega^{1+\delta}} d\omega = \frac{M_0}{\delta\omega_x^{\delta}} \tag{A.12}$$

So for any positive $\varepsilon > 0$, there exists a frequency ω_x such that the integral of the log sensitivity function over frequencies higher than this is bounded by this value. So if $T(s)$ has no imaginary poles then,

$$lim_{x\to\infty} \int_0^{\omega_x} \log|S(j\omega)| \frac{x^2}{x^2 + \omega^2} d\omega = \int_0^{\omega_x} \log|S(j\omega)| d\omega \tag{A.13}$$

Utilizing the decomposition of the sensitivity function using the Blaschke product, the following expression is obtained:

$$\log|\tilde{S}(x)| = \log|B_p^{-1}(x)| + \log|S(x)| = \sum_{i=1}^{N_p} \log\left|\frac{\overline{p}_i + x}{p_i - x}\right| + \log|S(x)| \tag{A.14}$$

Since the limit of $x \log|S(x)|$ is zero as $x \to \infty$,

$$lim_{x\to\infty} x\log|\tilde{S}(x)| = \sum_{i=1}^{N_p} lim_{x\to\infty} x\log\left|\frac{\overline{p}_i + x}{p_i - x}\right| \tag{A.15}$$

The following Maclaurin series expansion is employed for $|a| < 1$.

$$\log(1 + a) = a - \frac{a^2}{2} + \ldots + (-1)^n \frac{a^n}{n} + \ldots \tag{A.16}$$

So for $x > |p|$,

$$\log\frac{\overline{p}_i + x}{x - p_i} = \log\frac{1 + \frac{\overline{p}_i}{x}}{1 - \frac{p_i}{x}} = \frac{\overline{p}_i + p_i}{x} + HOT \tag{A.17}$$

Thus, $lim_{x\to\infty} x\log\left|\frac{\overline{p}_i + x}{p_i - x}\right| = 2Re[p_i]$ and

$$\pi \sum_{i=1}^{N_p} Re[p_i] = \int_0^{\infty} \log|S(j\omega)| d\omega \tag{A.18}$$

Equation (A.18) indicates that the plot of log$|S(j\omega)|$ versus radian frequency has a greater area of sensitivity increase than sensitivity decrease if the loop transmission function has ORHP poles. This difference is proportional to the sum of the real parts of the unstable open-loop poles. If there are no unstable open-loop poles ($N_p = 0$), these areas are equal. This is the original formula found by H. Bode.

This relationship gives great insight into the limitations of control performance for systems with greater than first-order roll-off. It is noted that feedback systems with low-pass loop transmission functions that roll off at first order have no positive feedback (the T-plane plot lies in the first and fourth quadrants). For stable systems with greater than first-order roll-off (quite typical for actual systems where the response drops sharply at high frequency), the integral of log sensitivity with positive feedback is equal to that of negative feedback. As negative feedback is increased over an interval of frequencies, positive feedback is increased at other frequencies. From a design perspective, the compensator must be designed so that there is positive feedback at frequencies where there is low disturbance power.

Bibliography

1. Dorf, R. and Bishop, R., *Modern Control Systems*, 12th edn., PTR Prentice Hall, Upper Saddle River, NJ, 2011
2. Kuo, B. and Golnaraghi, F., *Automatic Control Systems*, Wiley, NJ, 2003
3. Chen, C.-T., *Linear System Theory and Design*, Oxford University Press, Oxford, UK, 1999
4. Antsaklis, P. and Michel, A., *A Linear Systems Primer*, Birkhauser, Berlin, Germany, 2007
5. Deakin, M.A.B., 'The development of the Laplace transform', *Archive for the History of the Exact Sciences*, vol. 25, no. 4, pp. 343–390, 1981
6. Grattan-Guinness, I., 'Laplace's integral solutions to partial differential equations', in Gillispie, C.C., *Pierre Simon Laplace 1749–1827: A Life in Exact Science*, Princeton University Press, Princeton, NJ, 1997
7. De Schutter, A.B., 'Minimal state-space realization in linear system theory: an overview', *Journal of Computational and Applied Mathematics, Special Issue on Numerical Analysis in the 20th Century (Vol. I: Approximation Theory)*, vol. 121, no. 12, pp. 331–354, Sep 2000
8. Horn, R.A. and Johnson, C.R., *Matrix Analysis*, Cambridge University Press, Cambridge, 1985
9. Ljung, L., *System Identification - Theory For the User*, 2nd edn., PTR Prentice Hall, Upper Saddle River, NJ, 1999
10. Lurie, B. and Enright, P., *Classical Feedback Control*, Marcel Dekker, New York, NY, 2000
11. Lurie, B., *Feedback Maximization*, Artech House, Dedham, MA, 1986
12. Bode, H., *Network Analysis and Feedback Filter Design*, D. Van Nostrand Company, New York, NY, 1945
13. Franklin, G.J., Powell, D., and Abbas Emami-Naeini., *Feedback Control of Dynamic Systems*, 6th edn., Pearson Higher Education, Inc. Limited, 2010
14. Astrom, K.J. and Hagglund, T., *PID Controllers: Theory, Design and Tuning*, 2nd edn., ISA: The Instrumentation, Systems, and Automation Society, Research Triangle Park, NC 27709, 1995
15. Astrom, K.J., 'Model uncertainty and robust control', *Lecture Notes on Iterative Identification and Control Design,* Lund, Sweden, pp. 63–100, Jan 2000
16. Freudenberg, J. and Looze, D., 'Right half plane poles and zeros and design tradeoffs in feedback systems', *IEEE Transactions on Automatic Control*, vol. 30, no. 6, pp. 555–565, June 1985

17. Braslavsky, J., Middleton, R. and Freudenberg, J., 'Performance limitations in a class of single-input two-output nonlinear systems', *Proceedings of the American Control Conference,* San Diego, CA, June 1999
18. Astrom, K.J. and Murray, R., *Feedback Systems: An Introduction for Scientists and Engineers*, Princeton University Press, Princeton, NJ, 2008
19. Sandberg, H. and Bernhardsson, B., 'A Bode sensitivity integral for linear time-periodic systems', *IEEE Transactions on Automatic Control*, vol. 50, no. 12, pp. 2034–2039, 2005
20. Middleton, R., 'Trade-offs in linear control system design', *Automatica*, vol. 27, no. 2, pp. 281–292, March 1991
21. Kirk, D., *Optimal Control Theory: An Introduction*, Prentice-Hall, New Jersey, NJ, 1970
22. Horowitz, I., 'Quantitative feedback theory', *IEEE Proceedings on Control Theory and Applications*, vol. 129, no. 6, pp. 215–226, 1982
23. Hu, T. and Lin, Z., *Control Systems with Actuator Saturation: Analysis and Design*, Birkhauser, Boston, MA, 2001
24. Kapila, V. and Grigoriadis, K., *Actuator Saturation Control*, Marcel Dekker, New York, NY, 2002
25. Sontag, E.D., *Mathematical Control Theory: Deterministic Finite Dimensional Systems*, 2nd edn., Springer, New York, NY, 1998
26. Tanaka, K., Ikeda, T., and Wang, H.O., 'Robust stabilization of a class of uncertain nonlinear systems via fuzzy control: quadratic stabilizability, H∞ control theory, and linear matrix inequalities', *IEEE Transactions on Fuzzy Systems*, vol. 4, no. 1, pp. 1–13, Feb 1996
27. Ho, D.W.C, Lu, G., and Zheng, Y., 'Global stabilisation for bilinear systems with time delay', *IEEE Proceedings on Control Theory and Applications*, vol. 149, no. 1, pp. 89–94, Nov 2002
28. Fridman, E. and Shaked, U., 'An improved stabilization method for linear time-delay systems', *IEEE Transactions on Automatic Control*, vol. 47, no. 11, pp. 1931–1937, Nov 2002
29. Meckl, P.H. and Seering, W.P. 'Feedforward control techniques achieve fast settling time in robots', *Automatic Control Conference Proceedings*, Evanston, IL, pp. 58–64, 1986
30. Balas, M.J., 'Feedback control of flexible systems', *IEEE Transactions on Automatic Control*, vol. AC-23, no.4, pp. 673–679, Aug 1978
31. Horowitz, I., *Synthesis of Feedback Systems*, Academic Press, New York, 1963
32. Horowitz, I. and Sidi, M., 'Synthesis of feedback systems with large plant ignorance for prescribed time-domain tolerances', *International Journal of Control*, vol. 16, no. 2, pp. 287–309, 1972
33. Skogestad, S. and Postlewaite, I., *Multivariable Feedback Control: Analysis and Design*, Wiley, NJ, 2005
34. Jiguan Lin Varaiya, P., 'Bounded-input bounded-output stability of nonlinear time-varying discrete control systems', *IEEE Transactions on Automatic Control*, vol. 12, no. 4, pp. 423–427, Aug 1967

35. Joshi, S.M. and Gupta, S., 'On a class of marginally stable positive-real systems', *IEEE Transactions on Automatic Control*, vol. 41, no. 1, pp. 152–155, Jan 1996
36. Zhou, K. and Doyle, J., *Robust and Optimal Control*, PTR Prentice Hall, Upper Saddle River, NJ, 1996
37. Antsaklis, P. and Sain, M., 'Feedback controller parameterizations – finite hidden modes and causality', in *Multivariable Control: New Concepts and Tools (A85-48776 24-63)*, D. Reidel Publishing Co., Dordrecht, The Netherlands, pp. 85–104, 1984
38. Wang, Q.-G., Lee T.-H., and He, J.-B. 'Internal stability of interconnected systems', *IEEE Transactions on Automatic Control*, vol. 44, no. 3, pp. 593–596, Mar 1999
39. Maciejowski, J.M., *Multivariable Feedback Design*, Addison-Wesley, Reading, MA, 1989
40. Safonov, M. and Athans, M., 'A multiloop generalization of the circle criterion for stability margin analysis', *IEEE Transactions on Automatic Control*, vol. 26, no. 2, pp. 415–422, Apr 1981
41. Smith, M., 'Generalized Nyquist stability criterion', *International Journal of Control*, vol. 34, no. 5, pp. 885–920, 1981
42. Macagnano, D. and de Abreu, G.T.F., 'Gershgorin analysis of random Gramian matrices with application to MDS tracking', *IEEE Transactions on Signal Processing*, vol. 59, no. 4, pp. 1785–1800, April 2011
43. Letov, A.M., *Stability of Nonlinear Control Systems*, Gostekhizdat, Moscow, 1955
44. Khalil, H., *Nonlinear Systems*, PTR Prentice Hall, Upper Saddle River, NJ, 1996
45. Murray, R., Li, Z., and Sastry, S., *A Mathematical Introduction to Robotic Manipulation*, CRC Press, New York, NY, 1994
46. Slotine, J.-J. and Weiping, L., *Applied Nonlinear Control*, PTR Prentice Hall, Upper Saddle River, NJ, 1991
47. Lur'e, A. I. and Postnikov, V.N., 'On the theory of stability of control systems', *Applied mathematics and mechanics*, vol. 8, no. 3, pp. 246–248, 1944
48. Vidyasagar, M., *Nonlinear Systems Analysis*, 2nd edn., Prentice Hall, Englewood Cliffs, New Jersey, 2002
49. Lin, G., Balachandran, B., and Abed, E.H., 'Absolute stability of second-order systems with asymmetric sector boundaries', *IEEE Transactions on Automatic Control*, vol. 55, no. 2, pp. 458–463, Feb 2010
50. O'Brien, J., *Feasible Solutions to Unstable Singularity in Parallel Robots*, PhD Dissertation, Rensselaer Polytechnic Institute, May 2001
51. Kim, J., Park, F.C., and Lee, J.M., 'A new parallel mechanism machine tool capable of five-face machining', *CIRP Annals*, vol. 48, no. 1, pp. 337–340, 1999
52. Megretski, A., 'Necessary and sufficient conditions of stability: A multiloop generalization of the circle criterion', *IEEE Transactions on Automatic Control*, vol. 38, no. 5, pp. 753–756, May 1993
53. William, N.F. and Mark N.L., 'Phase vs. Gain Stabilization of Structural Feedback Oscillations in Homing Missile Autopilots', *Proceedings of American Control Conference*, Evanston, IL, pp. 323–329, June 1985

54. Lurie, B., 'The absolutely stable Nyquist-stable non-linear feedback system design', *International Journal of Control*, vol. 40, no. 6, pp. 1119–1130, Dec 1984
55. O'Brien, J.F. and Neat, G.W., 'Micro-precision interferometer: pointing control system', *Proceedings of the 4th IEEE Conference on Control Applications*, pp. 464–469, Albany, NY, September 1995
56. Neat, G.W. and O'Brien, J.F., 'Micro-precision interferometer: fringe tracker control system', *Proceedings of 19th Annual Guidance and Control Conference*, Breckenridge, CO, February 1996
57. Neat, G.W., O'Brien, J.F., Nerheim, N., Calvet, R., Singh, H. and S. Shaklan., 'Micro-precision interferometer: first stabilized fringes', *Proceedings of the SPIE International Symposium on AeroSense, Conference on Spaceborne Interferometry*, vol. 1947, Orlando, FL, April 1995
58. Lurie, B. 'Nonlinear correction for feedback maximization, describing function approach', *American Control Conference*, pp. 1504–1507, San Diego, CA, June 1984
59. Lurie, B. 'Absolutely stable feedback system with dynamic nonlinear corrector', *Proceedings of the IEEE*, vol. 70, no. 8, pp. 869–870, 1982
60. Lurie , B. and Hadaegh, F., 'Control of systems with tiered actuators with application to interferometer optical delay line control', *11th International Conference on Integrated Navigation Systems*, St. Petersburg, Russia, May 1, 2003
61. Chu, C.-C., O'Brien, J.F., and Lurie, B.J., 'System identification and structural control on the JPL phase B testbed', *5th NASA/DoD CSI Technology Conference*, Lake Tahoe, NV, March 1992
62. Freudenberg, J. and Middleton, R., 'Properties of single input, two-output feedback systems', *International Journal of Control*, vol. 72, no. 16, pp. 1446–1465, 1999
63. O'Brien, J.F., 'Multi-path nonlinear dynamic compensation for rudder roll stabilization', *Control Engineering Practice*, vol. 17, no. 12, pp. 1405–1414, December 2009
64. Wu, F., Grigoriadis, K., and Packard, A., 'Anti-windup controller design using linear parameter-varying control methods', *International Journal of Control*, vol. 73, no. 12, pp. 1104–1114, Aug 2000
65. Carruthers, D.J., O'Brien, J.F., McInroy, J.E., and Yang, Y., 'Development of a voice coil-actuated limited-DOF parallel mechanism for vibration suppression', *Journal of Systems and Control Engineering*, February 2009
66. O'Brien, J.F., 'Trends in region 3 control system performance as wind turbine size increases', *Journal of Automatic Control and System Engineering*, conditionally accepted, September 2011
67. Nelson, J., Carruthers, D., O'Brien, J., and McInroy J., 'Design and control of a 2-DOF parallel mechanism using prismatic and revolute actuators', *IASTED International Conference on Robotics and Applications*, Cambridge, MA, November 2010

68. Chen, D. and Seborg, D.E., 'Multiloop PI/PID controller design based on Gershgorin bands', *IEEE Proceedings - Control Theory and Applications*, vol. 149, no. 1, Jan 2002
69. *20% Wind Energy by 2030: Increasing Wind Energy's Contribution to U.S. Electricity Supply*, U.S. Department of Energy, Energy Efficiency and Renewable Energy, July 2008
70. Fingersh, L., Hand, M., and Laxson, A., 'Wind turbine design cost and scaling model', Technical Report NREL/TP-500-40566, December 2006
71. Jonkman, J. and Buhl, M., 'FAST user's guide', National Renewable Energy Laboratory, NREL/EL-500-38230, Golden, CO, 2005
72. Van Amerongen, J., *Adaptive Steering of Ships-A Model Reference Approach to Improved Maneuvering and Economical Course Keeping*, PhD Thesis, Delft University of Technology, The Netherlands, 1982
73. Pierson, W.J. and Moskowitz, L., 'A proposed spectral form for fully developed wind seas based on the similarity theory of S.A. Kitaigorodsku', *US Naval Oceanographic Office Contract 62306–1042*, 1963
74. Oda, H., Ohtsu, K., Sasaki, M., Seki, Y., and Hotta, T., 'Rudder roll stabilization control system through multivariable auto regressive model', *Proceedings of the 3rd IFAC Workshop on Control Applications in Marine Systems*, Genova, Italy, 1992
75. Fossen, T., *Guidance and Control of Ocean Vehicles*, Wiley and Sons, New York, NY, 1994
76. Van der Klugt, P., *Rudder Roll Stabilization*, PhD Thesis, Delft University of Technology, The Netherlands, 1987
77. Blakelock, J., *Automatic Control of Aircraft and Missiles*, Wiley - IEEE, New York, NY, 1991
78. Keel, L. and Bhattacharyya, S., 'Robust, fragile, or optimal?', *IEEE Transactions on Automatic Control*, vol. 42, no. 8, pp. 1098–1105, Aug 1997
79. King, B.B. and Yuh-Roung Ou, 'Nonlinear dynamic compensator design for flow control in a driven cavity', *Proceedings of the 34th IEEE Conference on Decision and Control*, vol.4, pp. 3741–3746, Dec 1995
80. Kapasouris, P., Athans, M., and Stein, G., 'Design of feedback control systems for stable plants with saturating actuators', *Proceedings of the 27th IEEE Conference on Decision and Control*, vol.1, no. 7–9, pp. 469–479, Dec 1988
81. O'Brien, J.F., 'High order control with nonlinear compensation for rudder roll stabilization', *ASNE Automation and Control Conference*, Biloxi, MS, December 2007
82. Carruthers, D. and O'Brien, J., 'Two degree-of-freedom parallel mechanisms for vibration suppression and tracking', *SPIE Symposium on Smart Structures and Materials*, San Diego, CA, March 2011
83. O'Brien, J.F., McInroy, J.E., Bodtke, D., Bruch, M., and Hamann, J.C., 'Lessons learned in nonlinear systems, flexible structures, and robotics through experiments on a 6-legged platform', *Proceedings of American Control Conference*, Philadelphia, PA, June 1998
84. Lewis, F. and Syrmos, V., *Optimal Control*, Wiley-Interscience, New York, 1995

Index

Page numbers followed by *f* and *t* denotes the figures and tables respectively.

absolute stability, 49, 84
actuator rate saturation, 145
actuators
  physical limits, 32, 43
  'smart', 109
anti-windup controller, 127–128, 145–147
  blade rate saturation, 146, 147*f*
  gain reduction of, 146
  high-speed shaft rate, 146, 147*f*
asymptotic stability, 52, 76
autonomous systems, 77

bandwidth
  functional, 28
  limitations, 31–38
BIBO stability: *see* bounded-output, bounded-input (BIBO) stability
black box model, 24
  with Fourier analyzer, 24, 25*f*
Blaschke product, 165–166, 168
Bode loop response, 93–106
  defined, 93
Bode phase/gain relationship, 13
Bode plot, 11
  of nonminimum phase system, 14*f*
Bode sensitivity integral, 31, 41, 91
  and Nyquist-stable system, 109
Bode's integral theorem, 167–169
Bode step, 93, 96*f*
  defined, 96
bounded-output, bounded-input (BIBO) stability
  asymptotic stability and, 52
  of impulse response, 49–50
  linear, time invariant (LTI) system with zero initial conditions for, 49
  of marginally stable system, 50–51
  of multiple-input, multiple-output (MIMO) system with transfer matrix, 52
  of SISO, LTI system, 49–50
  state equations, 51–52
  state space realizations of, 52–54
  sufficiency, 49–50

case studies
  ship's rudder, 115–119
  wind turbines, 98–106
Cauchy's Theorem, 54
causal system, 8
characteristic loci, 68
circle criterion, 84–85, 87
  for a SISO system, 85, 86*f*
CLHP: *see* closed left half plane (CLHP)
closed left half plane (CLHP), 8
closed-loop feedback system, 1, 2*f*
  sensitivity, effect of feedback on, 29–30

closed-loop stability, 54, 57, 58, 61
  sensitivity *S(s)* for, 167–168
closed right half plane (CRHP), 8, 67, 165–166
collective blade pitch rate saturation, 145
command feedforward system, 41–45
  block diagram of, 42*f*
  example of, 43, 44*f*–45*f*
  feedback bandwidth limitations and, 42–43
  input-output function of, 41–42
complementary sensitivity, defined, 30
controllability matrix, 19–20
convolution integral, 50
CRHP: *see* closed right half plane (CRHP)
critical point, 55, 58, 65
crossover
  shaping loop response above, 95–97
  shaping loop response below, 93–95, 94*f*–95*f*

design process, linear, 91–121
  Bode loop response, 93–106 (*see also* Bode loop response)
  Nyquist-stable system, 109, 110*f*
  overview, 91–92
  phase stabilization, 106–108
  single-input, single-output, 113–121 (*see also* Single-input, single-output (SISO))
  two-input, single-output, 109, 111–113, 111*f*, 112*f*–113*f* (*see also* two-input, single-output (TISO))
differential equation
  for second-order system, 16–17
disturbances
  rejection of, feedback control system and, 3, 3*f*

eigenvalue decomposition of matrix, 68–69
eigenvalues, matrix, 20–21
eigenvectors, matrix, 20–21
equivalent linear system, 144, 144*f*
exponential stability, 77

feedback, 27–38
  approximation of, 28
  control (*see* feedback control system)
  effect on sensitivity, 29–30
  large, 28–29, 32
  negative, 28, 31
  positive, 29, 31
feedback compensator, 1, 2*f*
feedback control system
  application, effects of, 1
  closed-loop, 1, 2*f*
  design of linear components, 91–121 (*see also* design process, linear)
  open-loop, 1, 2*f*
  and parameter variation, 3
  tracking (*see* Tracking, feedback system)
feedback systems, stability of, 54
  internal stability, 67, 67*f*
feedforward system, 41–48
  command, 41–45 (*see also* command feedforward system)
  defined, 41
  prefilter, 46–48, 46*f*–47*f*
Fourier analyzer, 24
  black box model with, 24, 25*f*
F-plane plot, 55
frequency domain analysis, 11
frequency-domain models, plant, 9–16
  Bode phase/gain relationship, 13
  frequency response, 11–12
  Laplace transform, 9–10
  nonminimum phase lag, 14–16

nonminimum phase system, 13
transfer function/transfer matrix, 10–11
frequency response, 11–12
of Nyquist plot, 55–56
frequency separation, 114
functional bandwidth, defined, 28

gain stabilization, 107
Gershgorin bands, 69
Gershgorin's Theorem, 49, 68–74
globally postive definite, 77
gray box model, 24

integrator windup, 128
internal stability, concept of, 49, 67–68
feedback system, 67, 67*f*
of RHP pole-zero cancelations, 67

Jacobian matrix, 23, 78
Jordan transformation, 21

Kalman-Yakubovich-Popov Lemma, 84
kinematic set point control system, 78–82
control laws, 80
effect of mechanism singularities on, 82
error function for rotational pose component, 79
examples, 80–82
initial and final configurations, 79
planar Stewart platform, 78*f*, 80, 81*f*
three-parameter representation of orientation, 79

Laplace transform(s), 9–10
inverse, 10*t*, 50–51
large feedback and, 28
negative feedback and, 28
operator, 1
properties of, 10–11, 10*t*
of state differential equation, 51, 53
of state equation, 21
of system's response to disturbance signal $d(s)$, 165
large feedback, 32
defined, 28–29
left half plane (LHP) poles, 51
LHP poles: *see* left half plane (LHP) poles
linear, time invariant (LTI) system, 8, 13, 83
in feedback connection with a nonlinear system, 83*f*
with zero initial conditions for BIBO stability, 49
linear, time invariant single-input, single-output (LTI SISO), 106
linearization, 22–23
example, 23
linear system, 8
with transfer function, response of, 51
locally negative definite (decrescent) function, 77
locally positive definite function, 77
logarithmic gain, 11–12
loop response, shaping
above crossover, 95–97
below crossover, 93–95, 94*f*–95*f*
case study, 98–106
complete, 97–98, 98*f*
loop transmission function, 27, 46
Bode's integral theorem and, 167–169
with first-order roll-off, 91, 92*f*
of Nyquist Criterion, 61–62
Nyquist plot of, 58–59
with second-order roll-off, 91, 92*f*
TISO, 111
T-plane plot of, 55–57
loop transmission matrix, internal stability of, 68
Gershgorin bands of, 69

LTI SISO: *see* linear, time invariant single-input, single-output (LTI SISO)
LTI system: *see* linear, time invariant (LTI) system
Lure-type Lyapunov function, 86
Lyapunov function, 80
  derivative of, 85
  quadratic symmetric, 84
Lyapunov's direct method, 77
Lyapunov stability theory, 49, 74–77
  local stability, 77
  uniformly globally asymptotically stable, 77
  uniformly locally asymptotically stable, 77
  uniformly locally stability, 77

Maclaurin series expansion, 168
marginally stable system, 50–51
marginal stability, 52
mass-spring-damper system, 7–8
mathematical model, of plant dynamics, 7–8
memoryless nonlinearity, 83
MIMO: *see* multiple-input, multiple-output (MIMO)
minimal state space realization
  defined, 20
minimum phase (MP) system, 13
MP system: *see* minimum phase (MP) system
multiple-input, multiple-output (MIMO), 8, 52

NDC: *see* nonlinear dynamic compensator (NDC)
negative feedback, 31
  defined, 28
NMP system: *see* nonminimum phase (NMP) system
nonlinear dynamic compensator (NDC), 128–129, 128*f*–129*f*
  anti-windup controller, 145–147
  for multiple saturations, 148–155
  multipurpose, 143–155
  SITO feedback system with, 143*f*–144*f*
  for a vibration suppression system, 129–142
nonminimum phase lag, 14–16
nonminimum phase (NMP) system, 13
  Bode plot of, 14*f*
  delay, 15–16, 16*f*
notch, 46
Nyquist contour, 54–55
  modified, 60–61
Nyquist plot, 55, 56*f*–63*f*
  for absolute stability analysis, 137*f*
  of closed-loop system, 61–62, 63*f*
  of $C(s)P(s)$, 86*f*
  Gershgorin Circles on, 75*f*
  of plant, 107, 107*f*
  for T_{ASFG}, 136, 137*f*
  of *T*-plane, 62–63
Nyquist Stability Criterion, 49, 55–63
  generalized, 68
  loop transmission function, 61–62
Nyquist-stable controller, 67, 128
  design, 134, 135*f*
  equivalent systems for, 138*f*
  Nichols plots for, 133*f*, 136
  practical applications, 141
  unstable nature, 141, 142*f*
Nyquist-stable system, 109, 110*f*
  controller, 120–121

observability matrix, 20
OLHP: *see* open left half plane (OLHP)
open left half plane (OLHP), 8
  poles, 50
open-loop feedback system, 1, 2*f*
  weaknesses in, 2

open-loop gain, 27
open-loop stability, 85
open right half plane (ORHP), 8
  poles, 55
    Blaschke product of, 165
    constraints of $S(s)$ at, 166
ORHP: *see* open right half plane (ORHP)
overshoot
  excessive, 4*f*
  as transient response characteristic, 4

parallel robots, multiaxis control of, 69–74, 70*f*
  compensator design, 71–74, 73*f*–74*f*
  end-effector, 70
  fourth-order compensators, 71
  frequency responses, 71, 71*f*–74*f*
  limited-DOF parallel mechanism, 70
  passive joints, 70
  plant identification, 70–71, 71*f*–72*f*
  set point control of, 78
parallel robots, *PUS–RR* model, 129, 130*f*
parameters
  variations, feedback control system and, 3
phase margin, of *T*-plane plot, 64
PID: *see* proportional-integral-derivative (PID) controller
plant, 1, 2*f*
  alignment, 114
  characteristics, 7
  descriptions, 7–25
  dynamics, mathematical model of, 7–8
  frequency-domain models, 9–16 (*see also* Frequency-domain models, plant)
  knowledge, 37, 43
  limits
    poles, 32–34, 33*f*, 43
    zeros, 34–37, 35*f*–37*f*, 43
  linearization, 22–23
  Nyquist plot of, 107, 107*f*
  sensitivity, 38*f*
  time-domain models, 16–22
  tracking (*see* Tracking, feedback system)
Poisson integrals, 166
pole shifting, 85
Popov Criterion, 86–87
  absolute stability and, 136–137
positive feedback, 31
  defined, 29
prefilters, 4, 46–48, 46*f*–47*f*
proportional-integral-derivative (PID) controller, 127
proportional-integral (PI) compensator ($C(s)$), 127, 128*f*

quadratic symmetric Lyapunov function, 84

ramp function, 51
relative stability, 63–67
return difference, 27
return ratio, 27
  SITO, 119*f*
RHP zeros: *see* right half plane (RHP) zeros
right half plane (RHP)
  poles, 51, 54
  zero natural frequency, 59
  zero phase, 59
right half plane (RHP) zeros, 12
  example, 14–15
  impact on feedback controller, 37
rise time, as transient response characteristic, 4
Routh Criterion, 54

rudder, ship
  case study, 115–119
  mathematical model of, 118, 118*f*

scalar gain, 55
SDRS: *see* Standard deviation to rated speed ratio (SDRS)
second-order system, 7–8
  differential equation for, 16–17
sector condition, 83
sensitivity
  complementary, 30
  defined, 30
  effect of feedback on, 29–30
  plot, 38*f*
sensitivity function, 167–168
sensor noise, 95
  as bandwidth limitation, 31–32
Seoul National University's Eclipse 5-face machining robot, 79*f*
  Eclipse manipulator, 80–82
settling time
  excessive, 5*f*
  as transient response characteristic, 4
signum function, 50
single-input, single-output (SISO), 8, 49
  block diagram of, 27, 27*f*
single-input, two-output (SITO), 8, 113–121
  block diagram of, 113*f*
  closed-loop step response, 116*f*
  control, example of, 115
  return ratio, 119*f*
SISO: *see* single-input, single-output (SISO)
SITO: *see* single-input, two-output (SITO)
SITO feedback system, 143*f*–144*f*
  FAST simulations, 156, 157*f*
  high-speed shaft rate, 156, 157*f*, 158, 159*f*
  HSS rate error variable gain, 156–160, 158*t*
  tower fore/aft acceleration, 160*f*
  variable gain for, 155–160, 158*f*, 158*t*
'smart' actuators, 109
s-plane contour, 59–62
Standard deviation to rated speed ratio (SDRS), 121
state, defined, 16–17
state differential equation
  solution of, 17–18, 19*f*
state matrix
  diagonalization of, 20–21
state space realization
  controllability, 19–20
  minimal, 20
  observability, 20
state space realizations, of BIBO stability, 52–53

Taylor expansion, 22
TFAAR: *see* tower fore/aft acceleration ratio (TFAAR)
time-domain models
  plant, 16–22
time invariant system, 8
TISO: *see* two-input, single-output (TISO)
tower fore/aft acceleration ratio (TFAAR), 156, 160
T-plane plot, 55, 107, 108*f*
  gain margin, 65, 65*f*–66*f*
  of loop transmission function, 55–57
  Nyquist plot, 62–63
  of Nyquist-stable system, 109, 110*f*
  phase margin, 64
  in phase-stabilized controllers, 65–66, 65*f*–66*f*
  zero net encirclements, 63–64

tracking, feedback system, 2–5, 2*f*
  disturbances rejection, 3, 3*f*
  sensitivity to parameter variation, 3
  transient response, 4–5, 4*f*–5*f*
transfer function, 10–11
  from state equation, 21–22
transfer matrix, 11
  stability of, 67
transient response, 4–5, 4*f*–5*f*
Tustin's method, 140
two-input, single-output (TISO), 8, 109, 111–113, 111*f*, 112*f*–113*f*
  block diagram, 111*f*
  controller, example of, 112, 112*f*–113*f*
  step response of, 113*f*

unit step function, 51
unmanipulable singularity, 82
unstable singularity, 82

vibration suppression system, NDC of, 129–142
  absolutely stable fixed gain (ASFG) controller, 132–134
  absolute stability analysis, 136, 136*f*–137*f*
  actuator, 130
  closed-loop performance, 140–141, 142*f*
  controller implementation, 139–140
  disturbance signals, 140, 141*f*
  gain zeros and poles of $P(s)$, 131
  gyroscope sensors, 130–131, 131*f*
  loop transformations, 138*f*
  nonlinear dynamic compensator (NDC), 137–139, 139*f*
  Nyquist-stable (NS) controller $C_{NS}(s)$, 134, 135*f*
  pole-zero-gain (PZK) model of, 131, 132*f*
  Popov Criterion, 136–137
  saturation mechanisms, 134–136
  system identification, 131–132

white box model, 24
wind turbines, case study, 98–106

zero input stability, 52–54
zero-state response, 50

www.ingramcontent.com/pod-product-compliance
Lightning Source LLC
LaVergne TN
LVHW011803200926
848484LV00003B/775
9781849194815